KU-105-892

New Techniques in Cattle Production

New Techniques in Cattle Production

Edited by
C. J. C. Phillips
School of Agricultural and Forest Sciences,
University College of North Wales, Bangor, UK

Butterworths
London   Boston   Singapore   Sydney   Toronto   Wellington

All rights reserved. No part of this publication may be reproduced in any material form (including photocopying or storing it in any medium by electronic means and whether or not transiently or incidentally to some other use of this publication) without the written permission of the copyright owner except in accordance with the provisions of the Copyright, Designs and Patents Act 1988 or under the terms of a licence issued by the Copyright Licensing Agency Ltd, 33–34 Alfred Place, London, England WC1E 7DP. Applications for the copyright owner's written permission to reproduce any part of this publication should be addressed to the Publishers.

Warning: The doing of an unauthorised act in relation to a copyright work may result in both a civil claim for damages and criminal prosecution.

This book is sold subject to the Standard Conditions of Sale of Net Books and may not be re-sold in the UK below the net price given by the Publishers in their current price list.

First published 1989

© Butterworth & Co. (Publishers) Ltd, 1989

British Library Cataloguing in Publication Data

New techniques in cattle production
  1. Livestock: Cattle. Production
  I. Phillips, C. J. C.
  636.2′08
  ISBN 0-408-03321-5

Library of Congress Cataloging in Publication Data

New techniques in cattle production/edited by C. J. C. Phillips.
    p.   cm.
    "Proceedings of the Second International Symposium
on New Techniques in Agriculture, held at the
University of Wales, Bangor, UK, in September,
1988" — Pref.
    Bibliography: p.
    Includes index.
    ISBN 0-408-03321-5 :
    1. Cattle—Technological innovations—Congresses.
I. Phillips, C. J. C.   II. International Symposium on
New Techniques in Agriculture (2nd : 1988 : University
of Wales)
SF191.2.N49   1989
636.2—dc20

Composition by Genesis Typesetting, Borough Green, Sevenoaks, Kent
Printed and bound by Hartnolls Ltd, Bodmin, Cornwall

Preface

This book is the Proceedings of the Second International Symposium on New Techniques in Agriculture, held at the University of Wales, Bangor, UK in September, 1988. The aim of the book is to provide an overview of research developments in the field of cattle production that will be of use to students, extension workers, researchers and leading farmers. It is hoped in particular that a multidisciplinary approach to cattle production developments has been achieved. Indeed, it was common at the Symposium to hear cattle nutritionists comment on how much they had enjoyed the papers on cattle reproduction, or mechanization, and vice versa. By considering cattle production development in a wider context than just the effect on productivity in a particular area, new techniques and technology can be more rapidly and effectively applied. In particular it is now increasingly realized that the effects of new technologies on the environment must be assessed.

We are not short of resources in cattle production: it is arguably the world's biggest industry, and the support industry in terms of research and extension workers, veterinarians and product processors is equally large when viewed internationally. There is therefore a need regularly to review the progress of the industry as a whole and to discuss the application of new techniques within the production sector.

New techniques do not necessarily equate with useful techniques in all circumstances, but in this context the authors of each chapter were asked to consider new techniques and technologies that had been developed in the last 10 years which were expected to reach the market place within the next 10 years. Some 'new' techniques may have been researched at an earlier date, but have remained dormant until the market place was right or the tools for applying them were available. For example the original research demonstrating the ability of anterior pituitary extract to increase milk yields in dairy cattle was carried out in the 1930s, but the technology to produce an analogue of the causative agent (growth hormone) at sufficiently low cost has only recently become available through genetic engineering. Short communications on the development of specific new techniques are included in Chapter 17.

In some fields, notably cattle nutrition, new techniques that are likely to have a major impact on farm practice are in remarkably short supply. In part this must reflect the depth of research that has already been conducted and applied in certain areas of cattle production. As the response to new research input dimishes in a certain field the role of the researcher often transfers to modelling the response patterns and integrating them with other disciplines, as is demonstrated in the chapter by Drs Bruce and Broadbent which is largely concerned with modelling the response of suckler cows to cold stress. The current growth area in cattle research is in the environment for cattle and several chapters in this book are devoted to this subject. Benefits of environmental research are not always strictly delineated in terms of cattle productivity, but may have implications for the comfort and welfare of cattle, the pollution of the environment and the acceptability of cattle production systems to an increasingly demanding public.

Of particular importance in any consideration of new techniques for cattle production is the incorporation of the techniques and technology into systems of production. This is true not only in the developed world where the traditional input:output evaluation of new techniques is now confounded by public concern for environmental and animal welfare issues, but also in developing countries where cattle are kept not just for food production but as producers of fuel and power, as symbols of social status and as a means of reducing risk in the use of land for productive purposes.

Finally, I am grateful to all those who made the Symposium, and I hope this book, a success: Professor I.F. Marai for suggesting the title of the Symposium and providing valuable advice; Arfon Borough Council, BOCM Silcock Ltd, Richard Keenon and Co. Ltd, and the University Federation for Animal Welfare for financial support; Professor J.B. Owen and Professor M. Aboul Ela for providing considerable help in the running of the Symposium and preparation of the book; Mrs P. Ellis, Mrs G. Owen and Miss G. Lee for dedicated secretarial work, Mr J.B. Thomas and Mr E.A. Williams for technical assistance and my wife for running the Accompanying Persons' Programme; to *Animal Production* for permission to reproduce the abstraction on p. 225; and finally to the authors of this book for rapidly and diligently preparing the manuscripts of the chapters that you are about to read.

C.J.C. Phillips
November 1988

Contributors

A.S. Abdel-Aziz, BSc, MSc, PhD
Professor, Faculty of Agriculture, University of Cario, Giza, Egypt

M.B. Aboul-Ela, BSc, MSc, PhD
Professor, Faculty of Agricultural Sciences, University of the UAE, PO Box 15551, Al Ain, United Arab Emirates

J.L. Albright, BSc, MSc, PhD
Professor, Department of Animal Sciences, Purdue University, West Lafayette, IN 47907, USA

P.J. Broadbent, BSc, PhD, CBiol, MIBiol
North of Scotland College of Agriculture, 581 King St, Aberdeen, UK

J.M. Bruce, MSc, PhD, CEng, MIMechE, MBIM
Centre for Rural Building, Craibstone, Bucksburn, Aberdeen, UK

A.S. El-Shobokshy, BSc, PhD
Professor, Department of Animal Production, University of Zagazig, Egypt

J.M. Forbes, BSc, PhD, DSc
Professor, Department of Animal Physiology and Nutrition, University of Leeds, Leeds, UK

J.F.D. Greenhalgh, MA, MS, PhD
Professor, School of Agriculture, University of Aberdeen, 581 King St, Aberdeen, UK

D.I.H. Jones, BSc, PhD
Welsh Plant Breeding Station, Plas Gogerddan, Bow Street, Aberystwyth, Dyfed, UK

M. Kelly, BSc, MSc, PhD
Buildings Department, West of Scotland College, Auchincruive, Ayr, UK

J.W.B. King, BSc, PhD
Professor, Edinburgh School of Agriculture, West Mains Road, Edinburgh, UK

J.D. Leaver, BSc, PhD, CBiol, FIBiol, FRAgS
Professor, Wye College, University of London, Ashford, Kent, UK

I.F.M. Marai, BSc, MSc, PhD
Professor, Faculty of Agriculture, University of Zagazig, Egypt

J.B. Owen, BSc, PhD, MA, FIBiol
Professor, School of Agricultural and Forest Sciences, University of Wales, Bangor, Gwynedd, UK

C.J.C. Phillips, BSc, PhD
School of Agricultural and Forest Sciences, University of Wales, Bangor, Gwynedd, UK

T.R. Preston, PhD, DSc
Convenio Interinstitucional para la Produccion Pecuria en al Valle del Rio Cauca, AA 7482, Cali, Colombia

J.F. Roche, MAgSc, PhD, DSc
Professor, Faculty of Veterinary Medicine, University College, Dublin, Ballsbridge, Dublin 4, Eire

D.W.B. Sainsbury, BSc, MA, MRCVS, PhD
Department of Clinical Veterinary Medicine, University of Cambridge, Madingley Road, Cambridge, UK

G. B. Scott, BSc, PhD
Buildings Department, West of Scotland College, Auchincruive, Ayr, UK

S.L. Spahr, BSc, PhD
Professor, Department of Animal Sciences, University of Illinois, Urbana, IL 61801, USA

W.R. Stricklin, BSc, MSc, PhD
Associate Professor, Department of Animal Sciences, University of Maryland, College Park MD 20742, USA

P.C. Thomas, BSc, PhD, CBiol, FIBiol
Professor, West of Scotland College, Auchincruive, Ayr, UK

L. Vaccaro, BSc, PhD
Professor, Instituto de Produccion Animal, Universidad Central de Venezuela, Maracay, Venezuela

A.J.F. Webster, MA, VetMB, MRCVS, PhD
Professor, School of Veterinary Sciences, University of Bristol, Langford, Bristol, UK

Contents

Introduction

J. M. D. Prescott*

The symposium upon which this book is based has addressed a remit of cattle production in a world-wide context featuring the opportunities and constraints that relate to both developed and developing countries. We shall all take away our personal impressions; these are mine, interrelated with my recent experience and opinions.

Professor Greenhalgh in chapter 1 notes that the striking feature of cattle production systems is their diversity. This has been well represented in other parts of the book.

In the next two chapters, contrasting approaches are sharply focused by Dr Preston and Professor Vaccaro on the one hand and Professor Spahr on the other in their thought provoking and informative papers.

Professor Greenhalgh asks – can such diversity be justified? Dr Preston and Professor Vaccaro emphasize the critical need in the developing countries to match production systems to available resources:

Step 1   identify the potential of available local resources
Step 2   match production to resources with integrated mixed farming systems.

In most of these countries cattle have a greater importance for draught, for fuel and for security than for milk or meat production. They state in their introduction 'even in those regions where famine appears to be endemic, attention must be focused on the overall issues of rural development and preservation of the environment, and not simply on how to increase supplies of milk and meat.' Employment and income generation not just food production are of concern.

They advance a persuasive justification for proper priority for established production systems combining calf suckling and hand milking in appropriate circumstances, because of the advantages in use of local feed resources and livestock health and in the gainful employment of people.

Professor Spahr provides a comprehensive review and update on developments in mechanization and automation which have resulted in dramatic reductions in the manpower required to produce milk in the USA and other developed countries. In relation to these he notes that there is multinational awareness and demand for this technology as it relates to feeding, milking and recording. He summarizes the state of development of sensors to monitor reproductive status and mastitis and the prospects for robotic milking. In his chapter he refers to advances in artificial

*Principal, Wye College, University of London, UK

intelligence that will enable the dairy man to interrogate computerized recording systems using natural language.

These two chapters certainly provide a marked contrast. But throughout the book it is evident that a diversity of cattle production systems is required to match the variety of physical, economic and social circumstances that prevail. Professor Greenhalgh exemplifies the major effects of economics by reference to the UK and New Zealand. Furthermore he suggests that cattle production even in the developing countries is likely to be 'demand led' and dependent on economic growth in these countries.

With the increased potency and precision of new techniques becoming available in many fields and the facility for them to be more widely applied, it is important that there should be renewed consideration not only of the means but also of the ends: i.e. the objectives to which new technology will be directed. Professor Greenhalgh provides a world perspective on this and made reference to both social and economic consideration.

Professors Aboul-Ela and Abdel-Aziz (chapter 16) deal with the incorporation of new techniques into cattle production systems in developing countries and Professor Leaver returns to this theme in chapter 15 relating to temperate conditions and developed economies.

It is my view that in Europe and North America cattle production will be required to meet increasingly specific consumer requirements, face intensified competition from pig and poultry meat and indeed vegetable protein, operate under the constraint of production quotas and reducing prices and adapt to increasing public concern about the influence of agricultural practices on the environment. All of these are mentioned in this book.

New techniques of cattle production will be required to:

improve efficiency of resource use
lower input costs
improve predictability of production
match production to more specific objectives with respect to milk protein and lean beef production
reconcile agricultural with environmental and amenity considerations.

In the UK, cattle production is likely to develop differently depending on the nature of the land and climate and the alternative options that are available for land use.

Most dairy farmers who remain in milk production are likely to be responsive to advances in technology including improved grasses and forage crops, more precise grazing management, improved nutritive value of silage and its strategic supplementation and enhanced milk production potential in the cows if these can be achieved by means acceptable to consumers! There will need to be a priority on more efficient use of fertilizers and better management of silage making and slurry disposal to protect more adequately the environment.

Lower input systems of cattle production on grass:clover swards are likely to become more widely adopted. Improved varieties of white clover and management to maintain its persistence and the productivity of grass:clover swards when they are managed for grazing and conservation will increase the reliability of clover-based systems. There is sound evidence that for beef systems this can reduce input of both concentrate feed and fertilizer and provide for more predictable production of finished beef cattle. I consider that there could also be prospects for lower cost and more extensive, but nonetheless profitable milk production from grass:clover swards provided the risk of bloat can be controlled, probably by set stocking and appropriate buffer feeding with silage. The prospective advantages in terms of low cost milk production could be considerable.

An increasing proportion of grassland in the UK is being designated as Environmentally Sensitive Areas or as Sites of Special Scientific Interest. In relation to these, farmers are invited or instructed to follow certain prescribed management practices which are intended to conserve species diversity in the flora and fauna and an attractive and varied landscape. The management guidelines are specific to each area or site but generally prohibit the use of inorganic fertilizer, herbicide and pesticide and permit only the use of grass for grazing or late cutting of hay. Farmers receive payments for accepting these management agreements and farming within these constraints. Low input cattle production as a means of vegetation management will complement such wildlife conservation and amenity use of land.

Overall future developments of cattle production will require a much more integrated approach not only in relation to agriculture but also alternative land use and the wider environment.

Advances in techniques

This book has brought together authoritative, critical and perceptive reviews of advances in science and technology. I wish selectively to highlight those features that appear to me to merit further emphasis and comment.

Better control of grazing management

This topic is critically reviewed by Dr Phillips (chapter 8). Significant progress has been made in the development of techniques that can be applied in the grazing situations to measure plant growth, grazing behaviour and forage intake (*see* Huckle, Clements and Penning, p. 236). The interaction between the growth of the sward and its utilization by cattle has become better understood and in particular the association between sward state (as indicated by leaf area index or surface height) and herbage growth, senescence and intake by grazing livestock has been quantified in field experiments and simulated by computer modelling. From this work have derived practicable guidelines for grazing management which can be applied to regulate grazing intensity at critical stages of the season without undue complexity of management or high cost of fencing. These guidelines are now being taken up and widely applied by development and advisory services. They facilitate the tactical adjustment of grazing and its integration with conservation using flexible set stocking and the strategic use of supplementary feeds, including rumen delivery devices for supply of trace nutrients and rumen modifiers to grazing cattle.

Improved feed processing, conservation and upgrading

El-Shobokshy *et al.* (chapter 6) review a range of new technologies that are now available to enhance the utilization of concentrate, byproduct and forage feeds and their presentation in mixed diets.

I wish to comment on two aspects; silage conservation and upgrading of low quality forage.

Silage currently provides over 66% of the total forage dry matter that is conserved in the UK. The technology for clamp silage making has advanced rapidly over the past decade. As grass silage has come to be widely adopted as the basis for winter feeding of most dairy herds and many beef units, so important limitations in its feeding value have become better recognized and concern about environmental pollution is intensifying. Current research is concerned with improving control of microbial fermentation in the silo, with particular interest centring on the critical features that influence the success of microbial inoculations and the concurrent use of enzyme additives to influence substrate supply. There are promising prospects that this approach will lead to the current range of corrosive chemical additives being substantially replaced by biological additives within the next few years.

Problems with silage effluent are a major preoccupation on many dairy and beef farms as legal actions relating to water pollution have increased. El-Shobokshy *et al.*, refer to the research in progress to reduce effluent loss by the incorporation of dry absorbent material such as rolled barley or dried beet pulp when direct cut grass is ensiled. The incorporation of rolled barley and dried sugar beet pulp has improved the fermentation quality of the silages which have supported high rates of gain (approaching 1 kg/head per day) in beef cattle. Alternative approaches involve the conservation for more mature and drier crops, including whole crop cereals, and their preservation and upgrading by ammoniation in clamp, stack or bale are being developed.

In developing countries the need to exploit more fully low quality forage is critical to draught power and much of crop production as well as to production of milk and meat. I share the enthusiasm of El-Shobokshy *et al.* for the application of relevant local technology, not involving agrochemicals, but including suitable biological techniques, possibly involving fungal cultures, for upgrading low quality forages. There is also a more general need pointed out by Preston and Vaccaro for better characterization of locally available forages and byproducts and their more effective incorporation in cattle diets in many of these situations.

More specific supplementary feeding and predictable production of milk and meat

Professor Thomas (chapter 7) describes the progressive improvement in our understanding of cattle nutrition as it relates to energy and protein in the dairy cow. He highlights current prospects for improved feed characterization and diet formulation; prediction of post-ruminal nutrient supply; and the prediction of production response in milk and meat output which requires the adequate characterization of animal potential.

He notes that the conceptual framework for further progress is being provided by modelling. He emphasizes that progress depends on building up nutritional information brick by brick. Past research has provided the means to make changes with quantitative precision, increased reliance on forage and increased flexibility to make optimal use of resources in feeding cattle. Still further demands will be made on this flexibility and adaptability in future if cattle nutrition is to keep pace with changes in feed resource availability and animal potential.

Improved utilization of forages associated with enhanced intake and nutrient supply will make a major contribution to efficient cattle production over the next decade. In particular, production of milk and meat from silage fed with a minimum of supplementary feed is likely to become more consistent and predictable.

New technologies are being developed that will enable the potential nutritive value of grazed and conserved forage, byproducts and concentrate feeds to be quantified by more comprehensive and reliable means. These will derive from a better understanding of cell wall chemistry and the specific carbohydrates and protein fractions of feeds (*see* chapter 17). In particular, the development and evaluation of near infrared (NIR) reflectance technology for forage analysis offers promise for more comprehensive and discriminating assay of feed stuffs. The paper by Dr Barber and his colleagues (p. 228) represents a most important step forward with NIR spectra related to composition with prediction errors that are remarkably low (standard error of prediction of organic matter digestibility 2.6). The prospect is that feed samples taken on the farm will be analysed within minutes, with the results available as a computer print-out and the information transformed into a series of alternative rations or feed budgets.

The objective of these developments will be to provide a basis for more reliable prediction of milk protein and milk fat output and lean beef production from grazed and conserved forages.

Environment, housing, health and welfare

Adaptation to environment

The adaptation of cattle and of production systems to tropical environments is of particular concern in many countries. Professors Marai and Forbes' chapter (chapter 9) is partly concerned with the prospects for selecting cattle and adapting management techniques that will be suitable for hot dry climates. It highlights the difficulty of reconciling potential for high production with conditions of heat stress and its amelioration by economic means unless there is adequate cool water and it is feasible to use high concentrate diets.

Dr Mittal's (p. 242) experience in the hot arid region of India has been that there is little justification for the introduction of European blood into native cattle breeds even with intensive management. Furthermore, his paper highlights the production advantages of an unusually small indigenous breed of cattle that performs particularly well under these harsh conditions to which it appears to be exceptionally well adapted.

Housing

Drs Kelly and Scott (chapter 10) in their consideration of developments in winter housing for cattle reflect increased concern for animal comfort as well as ease of operation, but with a critical regard to the relative costs of alternative systems adapted to particular farm conditions. There is a need to reconcile comfort, cleanliness, durability and cost.

They highlight the following features; the straw bedded yards are too expensive in the West of Britain and that Enka mats are preferred in terms of cow comfort but too expensive at £50/cow to be taken up on farms; that even improved feeders reducing feed wastage by 10% have not so far proved to be sufficiently attractive in practice; and, that sloped floor systems best match needs at an acceptable cost.

Health and welfare

Professor Webster's concern with welfare, health and productivity (chapter 13) is to minimize calf death and disease. He gives an elegant description of the pathogen – host interrelationships. He notes that although cleanliness of the air is improved by ventilation, it is not possible to overcome the deleterious effect of overstocking by increasing ventilation. He comments that the fully developed rumen is the best probiotic medium available.

Drs Kelly and Scott note that sound progress in building design has benefited from careful observation and investigation of cattle behaviour; e.g. the recognition of the importance of 'lunging space' in the design of cow cubicles. This is a theme extended in a subsequent chapter in which Professors Albright and Stricklin treat us to a fascinating review of behaviour studies and husbandry experience relating to dairy cows, heifers and calves in the USA (chapter 12). They provide intriguing evidence of the effect of rearing calves in isolation compared with competitive group rearing situations on their subsequent behaviour and performance in the milking herd. They emphasize the importance of the contact and communications between stock and stockmen/women, who need to be able to read the sign language of animal behaviour. They draw attention to the codification of evidence and experience on animal welfare in livestock units used for teaching and research in the USA.

Reproduction and genetic improvement

The overall efficiency of animal production depends crucially on efficient breeding management of flocks and herds and in any respect, more effective monitoring of reproductive state and indeed regulation of reproductive performance is now a realizable possibility.

Professor Roche (chapter 4) reviewing progress in control of cattle reproduction states that 'oestrous control has taken 20 years to develop to its present state – similar time scales are likely to apply to many of the techniques currently being proposed involving applications of embryo transfer'. The development of *in vitro* fertilization is likely to be 'a long hard road'.

His assessment is that there are currently no clear cut techniques for improving oestrous detection: no hormone treatments to enhance conception rate or reduce embryo mortality; but that the progesterone releasing device (PRID) is preferred to prostaglandin for oestrous control; and, there are several promising alternative options for production of twin calves in prospect.

Professor King (chapter 5) in his appraisal of advances in the genetic improvement of cattle draws attention to the fact that artificial insemination still remains underexploited. He highlights the opportunities for more intense bull selection and notes that embryo transfer is an established technique and embryo sexing is now available, but that reliable methods of the semen sexing are still not feasible.

He comments that the application of embryo transfer to conventional progeny test provides only a modest 10% improvement, but that when applied to dairy cows, multiple ovulation and embryo transfer (MOET) accelerates improvements by 30%. It shortens generation intervals and allows closer control of the selection herd, with recording of inputs as well as output, thus allowing selection for efficiency. It allows selection to be quite independent of variation in the management of commercial herds where prospectively the use of bovine somatotrophin (BST) may confound genetic selection. When applied to beef breeding, it may well be more difficult to recoup costs of MOET schemes. However, a sex-determined bred heifer system could give more than a 25% improvement over conventional systems. Cloning by nuclear transfer would give a 'quantum leap' in beef improvement.

The most encouraging prospects for practical application of twinning is provided by the experience of Drs Broadbent and Dolman (p. 238). Their assessment is carefully qualified but most promising. Twinning rate in pregnant cows approached 50% to one exposure and could be raised to 70% by repeated exposure.

Regulation of nutrient partition and production of milk and beef

Improved understanding of the intrinsic control and regulation of reproduction, lactation and growth in all species of farm livestock is currently advancing more rapidly than at any time in the last 50 years. It will enable the biological variation in such characteristics as feed intake, reproductive performance, milk yield and composition, growth rate and carcass leanness to be explained and better exploited. In part this knowledge will be used in more effective predictability of performance and consistency of product quality and to regulate production better to match market requirements.

Improved understanding of the endocrine control of metabolism will have a role to play in this respect. Controversy presently relates to growth hormone or more specifically BST in relation to milk production. Administration under appropriate conditions results in a prompt increase of about 20% in the milk yield per cow and an improvement of about 10% in efficiency of feed conversion. However, Professor Roche (chapter 4) states that growth hormone in cattle is 'not strongly anabolic'. Similarly substantial increases in growth rate, carcass leanness and efficiency of feed conversion have been shown in lambs but Professor Roche has evidence that β-agonists are more effective by reducing protein degradation. Professor Greenhalgh suggests that manipulation of protein turnover offers promising prospects.

The importance of the present BST technology, which is currently undergoing unprecedented testing worldwide, depends less upon whether or not this particular technology is used to regulate milk and meat production and more on the subsequent developments that it heralds. A number of alternative approaches to influencing the partition of nutrients between fat and muscle are being investigated and from these are likely to emerge, ethically more acceptable, immunization procedures that will produce a highly specific regulation of the animal's own endocrine system and nutrient metabolism. Professor Greenhalgh refers to this as 'switching' and in this respect the paper by Drs Moore, McLauchlan and colleagues (p. 241) on immunization against LHRH is a pilot study of such a technique.

Enhanced animal production potential and more precise control of production systems will place more specific demands on feeding systems in the future. Developments in both should be complementary in reducing input costs and improving product quality. However, the application of this technology also raises important issues concerning the acceptability of milk and meat to consumers and the socioeconomic implications for grassland farmers as the progressive increase in the production per animal accelerates.

Computer modelling of cattle production

I share the optimism of Dr Bruce that modelling will make an increasingly important contribution to both the integration of scientific knowledge and the facilitation of its application in practice; features that were well exemplified in his joint chapter with Dr Broadbent (chapter 14). Modelling is an excellent catalyst around which to focus interdisciplinary research and development.

Increasingly over the next few years more and more cattle farmers are likely to adopt the use of computers and receive through computer terminals advice from development and extension agencies and from consultants. Additionally such simulation modelling may be used to assess the implications for production of the adoption of new enterprises or novel production techniques. Used with care it is a potent technique of considerable potential value in developing as well as developed countries. There is, however, a critical need to identify sound and robust objectives not just for production but in relation to rural development and sustainable agriculture, compatible with environmental protection.

Concluding remarks

We all depend on the same science base from which to address the different opportunities and constraints in developed and developing countries. Joint involvement in research, development and education with academic links such as those between University College of North Wales and Zagazig University, Egypt, which have generated this Symposium will enable sound progress to be made.

As a consequence of the chronic balance of payment problems facing developing countries and the reduction in funding of research and development in developed countries, scientific education relating to cattle production will be at an increased premium. There is an urgent need for more effective coordination and multidisciplinary approaches to rural development, in which advances in cattle production will play a major role.

Chapter 1

Current systems of cattle production and the potential for change

J. F. D. Greenhalgh

The aim of this chapter is to conduct a broad survey of current systems of cattle production, in order to identify weaknesses that may be corrected by the application of new technology or by other means. Neither individual systems nor individual techniques will be discussed in detail, but special attention will be given to economic and social factors determining the current status and future development of cattle production systems (Crotty, 1980), which have an important – and often dominant – influence on the application of new techniques.

Table 1.1 Contribution of milk and beef to human diets

	World	*Africa*	*Europe*
Population (m)	4837	555	492
Cow's milk			
Production (tonnes × 10⁶)	458	11.7	184
Provision (per head per day)			
milk (g)	260	58	1025
protein (g)	8.6	1.9	33.8
energy (MJ)	0.71	0.16	2.79
Beef			
Production (tonnes × 10⁶)	46.1	3.14	11.0
Provision (per head per day)			
beef (g)	26.1	15.5	61.2
protein (g)	4.12	2.45	9.67
energy (MJ)	0.31	0.18	0.72
All foods			
Provision (per head per day)			
protein (g)	69.1	57.9	98.7
energy (MJ)	11.2	9.7	14.3

(Courtesy of FAO, 1985).

(Milk is assumed to provide 2.72 MJ/kg and 33 g protein/kg; for beef the corresponding figures are 11.7 MJ and 158 g protein (Paul and Southgate, 1978).

Cattle and man

The contributions of milk and meat to the diet of man, shown in Table 1.1, are based on the statistics of the Food and Agriculture Organisation (FAO). The accuracy of the statistics must be questionable, particularly with regard to products consumed on their farm of origin, but they are the best available. As a world average, cow's milk apparently provides about 12% of man's protein intake and 6% of energy intake. The contributions of beef are less than half those of milk. Separate figures for Africa and Europe demonstrate the large differences in the contributions of cattle to human diets in developing and developed countries. In Europe, where milk is used in animal as well as in human diets, it could provide up to one-third of man's protein intake and 20% of energy; in Africa – a continent with a large cattle population – beef provides only 4% of man's protein intake and 2% of energy. Figures for continents, and even for individual countries, hide differences within populations, e.g. between social classes. In the UK, dietary surveys show surprisingly small variations between social classes (defined by income) in the contribution of animal products to total energy and protein intakes (MAFF, 1987). This desirable state of affairs, however, is unlikely to apply in less affluent countries.

The ranges in consumption of cattle products serve to demonstrate the inessentiality of cattle as providers of food for man. Yet this conclusion must be qualified. In some parts of the world, cattle are an essential intermediary in the conversion of fibrous vegetation to food edible to man. Over a much wider area, cow's milk is a vital component of the diets of infants. Cattle are also a vital source of draught power in many countries, and it has been estimated that 200 m of the world's 1300 m cattle act as draught animals (Holmes, 1983). In the developed countries, however, the relatively large contribution made by cattle products to the diet of man could, if necessary, be replaced by crop products, or by diversion of crop products to other livestock. There are signs that both kinds of substitution are taking place, and that this is due in large measure to fears of the effects of fats in cattle products on human health.

The exploitation of cattle by man has been restricted by social factors, and these are still important, e.g. in India. As human foods of animal origin are generally more expensive sources of nutrients than are foods of plant origin, cattle production over a large part of the globe is restricted by economic factors; the consumer cannot afford cattle products, hence they are not produced. Richer countries have lifted economic restrictions, either by improving the economy as a whole, or by subsidizing animal agriculture. But what we now see in the developed countries is a re-emergence of social restrictions on animal production. These range from concerns over the welfare of intensively housed and managed livestock (less important for cattle) to the conviction of a minority that animals should not be eaten. Although it is easy to argue that milk and beef are unlikely to be ruled out of the diets of man, it should be borne in mind that most social reform begins as a minority view. (For example, one has to go back only 100–200 years to find a time when the idea that man should not exploit his fellow man through slavery was a minority point of view.) In the shorter term, however, we shall continue to consume cattle products but must expect increasing resistance to techniques that cause excessive modification of animals and their products, especially if the products are judged to be prejudicial to human health. Another conflict between techniques and social factors is likely to arise over herd size. In the UK, for example, the average size of dairy herds has increased from 27 to 65 cows in 20 years. Future developments such as automated milking are likely to favour even larger herds. Increasing the size, and hence reducing the number, of herds is perceived to be damaging to the social structure of agriculture because it prevents the small farmer from keeping cows. Moreover, a widening of the ratio of stock to stockmen is potentially detrimental to animal welfare.

World cattle production systems on a global scale

Over the past 20 years there have been substantial increases in total cattle numbers (+18%) and in the production of milk (+34%) and beef (+39%) (Table 1.2). However, over the same period, the human population has increased at a faster rate, and the provision of cattle products per head of the human population has declined. The figures in Table 1.2 for Africa and Europe show contrasting pictures; in Europe, the production of both milk and beef per consumer has increased by 30%, while in Africa production per consumer of milk has fallen by 15% and that of beef has risen by only 10%.

Table 1.2 Growth of human population, cattle numbers and cattle products

		World	Africa	Europe
(a) Absolute				
Human population (m)	1965	3334	317	445
	1985	4837	555	492
Cattle population (m)	1965	1074	141	122
	1985	1269	177	132
Milk (tonnes × 10^6/year)	1965	343	8.0	137
	1985	458	11.7	184
Beef (tonnes × 10^6/year)	1965	33.1	1.6	7.2
	1985	46.1	3.1	11.0
(b) Relative (kg/year)				
Milk/human being	1965	103	25	308
	1985	95	21	374
Beef/human being	1965	9.9	5.1	16.2
	1985	9.5	5.7	22.3
Milk/cattle	1965	319	57	1122
	1985	361	66	1394
Beef/cattle	1965	30.8	11.4	59.1
	1985	36.3	17.7	83.3

Courtesy of FAO, 1965, 1985.

The output of cattle products per head of the cattle population has increased slowly over the past 20 years, by less than 1%/year. These figures show also that although in Africa beef production per animal is rising quite rapidly, the 1985 values for both beef and milk are still far below their European equivalents. From these figures one can calculate that cattle in Africa use about 85% of their total metabolizable energy requirement for maintenance; for European cattle the corresponding value is about 55%. A further point to note is that the production of alternative meats to beef has risen more rapidly. For example, in 1965 world pig and poultry meat production amounted to 12.9 kg/head of human population, but by 1985 this had risen to 18.4 kg (compared with 9.9 and 9.5 kg for beef).

What lessons can be learned for the future from the analysis of global (or continental) cattle systems? In the developed countries the successful intensification of cattle production has led to increasing dependence on concentrated feeds and hence to relatively unsuccessful competition with pigs and poultry for meat production (milk gives a different story, because it has fewer competitive products). In these countries, although some 'de-intensification' is feasible (e.g. the reduction in concentrate usage for dairy cows after the introduction of milk quotas), future improvements in cattle production are envisaged mainly in terms of further increases in output per animal. Thus the case for the European 10 000 litre 'super cow' has been argued by De Boer (1987). Because the markets for cattle products are either fixed or increasing only slowly, it is recognized that increased production per animal must mean fewer animals.

In the developing countries the slow progress in the productivity of cattle systems over the past 20 years (hence the widening gap between developed and developing countries), may be seen either as an opportunity for improvement or as a portent of further slow progress in the future. The optimist will point to the fact that if an animal is using 85% of its feed energy for maintenance, an increase in energy intake of only 15% will be enough to double its production rate. The pessimist will point to the low availability of the concentrated feeds required to increase energy intake, and perhaps also to the better use of these scarce resources by other livestock. What must also be taken into account – by both optimist and pessimist – is the market for milk and beef in developing countries. A close relationship between national figures for income and animal protein consumption per head of population was demonstrated many years ago (Food and Agriculture Organisation, 1964). The relationship between income (expressed as gross national product per person) and animal productivity (exemplified by milk yield per cow) is less close (Figure 1.1),

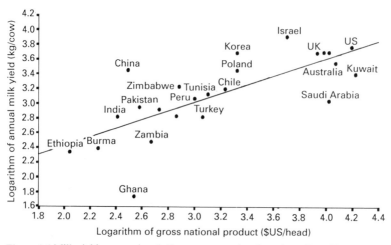

Figure 1.1 Milk yield per cow in relation to gross national product. (Stratified random sample of countries covering the global range in GNP). (Courtesy of FAO, 1985 and World Bank, 1986)

but gives a correlation coefficient (r) of 0.75. Crotty (1980), from a similar calculation, found $r = 0.52$, and reminds us that cause and effect may be confused. Nevertheless, it seems likely that it is consumers with money to spend who stimulate more efficient cattle production. For example, the recently expanded economies in Asia (Japan, Korea, perhaps now China, Taiwan etc.) now have milk yields per cow equal to or greater than many European countries. Another example is the growth of intensive milk production in the rich domestic markets of the Middle East oil states.

Diversity of cattle production systems

The most striking feature of cattle production systems is their diversity. Systems aimed at the same product (milk or meat) may use very different resources of cattle genotypes, feeds and other inputs. Table 1.3 compares two systems of milk

Table 1.3 Comparison of specialist, grass-based dairy systems in England and Wales and New Zealand

	England and Wales	New Zealand
Physical performance		
Farm size (ha)	79	64
Cow numbers	116	129
Milk (litres/cow)	5134	3270
(litres/ha)	7539	6590
Fat (kg/cow)	200	155
(kg/ha)	253	312
Financial information		
Milk price (p/l)	14.63	7.70
Milk sales (£/cow)	751	252
Feed variable costs (£/cow)	292	35
Employed labour (£/cow)	96	16
Interest charges (£/farm)	9995	4823
Profit (£/farm)	5739	8686

Sources: England and Wales, Milk Marketing Board (1985); New Zealand, New Zealand Dairy Board (1985).

£1 = 2.5 New Zealand dollars.

production practised in the UK and New Zealand. The two systems are based on similar breeds of cattle (mainly Friesians) and use pasture herbage as their basic input. The factor that drives the systems apart is the price paid to the producer for milk, this being almost twice as great in the UK as in New Zealand. Because of this, the New Zealand producer must use the cheapest possible inputs (no fertilizer N or expensive concentrates) and take advantage of a climate that provides a long grazing season and calls for minimal inputs of capital as buildings and machinery. An intelligent but naive observer of these systems might ask how they manage to coexist. Why does the New Zealand farmer not export his produce to the UK; why does the UK farmer not take more advantage of his environment to produce milk more cheaply? The answers to these questions are, of course, to be found in economics, not biology.

The New Zealand dairy industry can deliver butter to the UK at a much lower price than can the UK industry, but must then pay import duties to ensure that butter from the two sources has the same retail price. In the UK, and throughout the European Community, a guaranteed market for milk has encouraged the development of a capital-intensive dairy industry. Even the Republic of Ireland, with a grass-growing climate almost as good as that of New Zealand, has not chosen to develop a particularly low-cost dairy industry.

As with milk production, specialized beef production systems show great diversity across the world. Europe and North America produce beef at high cost; Australia and parts of South America operate lower cost systems. Elsewhere, beef production is frequently a byproduct from animals kept mainly for draught purposes and possibly also for milk. However, even within a single country there is diversity in beef production systems. In the UK, 60% of the beef produced is derived from the dairy industry, either directly (i.e. Friesian type cattle) or indirectly (beef × dairy crosses), and only 40% comes from pure beef breeds. Moreover, cattle of the same type (Friesian steers) may be finished for slaughter on diets ranging from all-cereals to all-forage. An observer from, say, manufacturing industry might marvel at the fact that such diverse production systems could be equally competitive. A poultry producer might contrast this situation with that of broiler production, for which a single system has been identified as optimal and is practised not only throughout the UK but also in many other countries.

Constraints on current systems

Jasiorowski and Quick (1987) divide the constraints on cattle production systems into those that are ecological, biological and socioeconomic. The ecological constraints – land and climate – are beyond the scope of this chapter. The socioeconomic constraints will be kept in mind, but the emphasis here is on the biological or technical constraints, of animal genotype, reproduction, nutrition and disease.

Animal genotype

The traditional methods of modifying the genotype of cattle have been selection within a defined population (i.e. increasing the frequency of desirable genes) and the introduction of foreign genes from other populations (i.e. breed substitution). To these measures can now be added the possibility of introducing specified genes (even synthetic genes) by recombinant DNA technology.

Breed substitution often forms the first phase of a cattle breeding programme. Its effects are rapid and dramatic; it can be very profitable for breeders supplying exotic stock. Selection within a breed should logically form the second phase of a programme. Its effects are achieved slowly and with less excitement. Many cattle breeding programmes have failed to move decisively into the second phase, but genetic improvement of dairy cattle in the developed countries is clearly concentrated on selection within a closely-related group of black and white breeds. In the UK, selection programmes for dairy cattle have been conducted largely by the milk marketing boards, but the rate of genetic improvement in milk yield in the national herd has been only about one-quarter of that theoretically possible (Nicholas and Smith, 1983). Yet there is evidence that herds which pursue a single objective by the best means available (i.e. inseminating cows only with semen from the best of progeny-tested bulls) can make progress at a rate close to the theoretical 2% improvement per year (Edinburgh School of Agriculture, 1988). Breeding schemes based on multiple ovulation and embryo transfer (MOET) now offer even faster rates of improvement (Woolliams and Smith, 1988).

Perhaps the major problem in dairy cattle breeding is not the means, but the end; in other words, the definition of selection objectives. In Europe and North America, the aim to produce higher yields of fat plus protein will surely have to be redirected towards protein. However, in developing countries the small amounts of milk available per human consumer may call for a different strategy for improving milk production. The need for dual (or triple) purpose cattle in the developing countries – which may arise again in countries with specialized dairying – must also complicate selection objectives.

Genetic improvement of beef cattle is far behind that of dairy cattle, as it is still dominated by breed substitution. The numerous breed societies have been slow to establish systematic selection programmes, but MOET schemes now offer them an opportunity to catch up (and at least one society in the UK has recognized this). The product to be maximized must be lean tissue, as fat can be added or subtracted by varying slaughter weight. In specialized (i.e. suckler cow) beef production systems the overhead cost of maintaining the cow is significant in both biological and economic terms. The efficiency of the suckler cow needs to be improved either by increasing output (*see below*) or by reducing inputs. It has long been recognized that for systems in which the female is kept almost entirely as a producer of offspring, a small female producing large offspring is required for maximum efficiency. Yet beef producers have been reluctant to accept this idea, partly because a special source of small females is required and partly because large calves sired by bulls of large-size breeds present calving problems. The idea of dimorphism remains attractive, however, and may in the future be made more practicable by the control of sex and possibly by exogenous control of gene expression.

Reproduction

A discussion of constraints on animal genotype progresses naturally to constraints on reproduction. The major constraint on reproduction in suckler cows is the limitation to an output of one calf (or fewer) per year. In dairy cows, however, the production of calves is sometimes seen as an unnecessary burden; ideally, the poorer cows would lactate continuously without becoming pregnant, the better cows would be bred to provide female offspring, and the best cows would provide bull calves. Techniques for the control of reproduction, especially MOET, are seen as the tools needed to remove the present constraints, but the effects of applying these techniques to cattle systems may not be those anticipated.

Embryo transfer offers the means by which suckler cows could produce twin calves, perhaps of a specified genotype, but the poor environments in which many suckler cows are kept (e.g. the rangelands of America and Australia, and even the hills of the UK) preclude twin production. In a country like the UK, with a cattle industry dominated by dairy production, embryo transfer may eventually be used to produce twin beef-type calves from the poorer dairy cows, rather than from suckler cows. (The technology needed to achieve this change is available today; all that might be needed to put it into effect would be a reduction in milk price and a switching of subsidies from suckler cows to an embryo supply service for dairy cows.) Further calculations show, however, that using the dairy herd to provide twin calves for beef production would not necessarily supplant the suckler cow. The number of dairy cows in the UK will decline as milk quotas fall and yields per cow rise, and if the number of cattle slaughtered for beef is to be maintained, an additional source of calves will be needed.

The suckler cow is therefore likely to maintain its position as a suitable case for technical improvement. Two alternatives to twinning have been suggested. The first (Taylor *et al.*, 1985) is a once-bred heifer system in which beef heifers produce their own female replacement (sex predetermined) and are then slaughtered. The advantage of this system is that animals do not have to be maintained for long periods of low productivity; the problem of instituting it in the UK would be to persuade the meat trade that once-bred heifers could provide high quality beef. The second alternative is to reduce the gestation period of cattle, so that calves could be produced at intervals of less than 12 months. Gestation is perhaps the only remaining period of life of domestic animals that has not been modified by the intervention of man. Fetal development seems to operate to a closely controlled time scale but the fetus is not a closed system (compare to the hen's egg) and is therefore vulnerable to external influence exerted through the dam by man. Certainly, the growth of a 40-kg calf in 280 days seems to be a leisurely rate of production when compared to the other production functions of cattle.

Nutrition

The genetic and reproductive constraints act mainly within the animal, and are relaxed by modifying the animal itself. Many (but not all) nutritional constraints can be tackled externally to the animal, most obviously by modifying feeds. The primary nutritional constraint on cattle nutrition is the rate at which the animal can extract energy-containing nutrients from the β-linked carbohydrate polymers that normally form a large part of its diet. Many traditional and new techniques for accelerating this process are known, the major ones being genetic improvement of forage plants, improved harvesting and conservation, feed processing, and supplementation. Strictly defined, supplementation means the correction of dietary deficiencies of protein, minerals and vitamins, but it also includes the provision of readily-digested carbohydrate sources that effectively reduce the load on cellulolytic microorganisms in the rumen.

If we accept that the major improvements in cattle productivity over the past 50 years are largely due to improved nutrition (and hence to relaxation of the constraints on fibre digestion), then we must also accept that on a global scale the most effective means of improving productivity has been the use in cattle diets of concentrates based on cereals and related materials. The relative success of cattle production in the developed countries has been achieved largely with concentrates, although improvements in forage quality and yield have also played a part. The relative failure of cattle production elsewhere has been due to lack of concentrates for cattle, and to the inability of other improvements in nutrition to compensate for inadequate 'supplementation' of forages. Future developments in the lifting of constraints are difficult to predict. In the developed countries, much research is still directed at the improvement of forages and their utilization, despite increasing surpluses of cereals. It is possible that genetic engineering will provide superior strains of bacteria with a voracious appetite for cellulose (Gregg *et al.*, 1987) but past experience indicates the difficulty of modifying the rumen ecosystem in a beneficial way. Genetic engineering may also yield better forage plants, but is likely to provide better cereals first.

Improvements in digestion in cattle need to be matched by, or supplemented with, improvements in metabolism. A major constraint on efficiency of metabolism is the cost of protein turnover, which affects requirements for both maintenance and production. Its effects can be demonstrated by comparing a cow producing 20 kg milk/day with a steer growing at 1 kg liveweight/day. Because the cow synthesizes protein that is not turned over, its net synthesis is over four times that of the steer (and it is also able to digest and metabolize 60% more feed). Growth promoters now under development (somatotrophin, β-agonists) influence protein turnover and the partition of nutrients, but it is possible that in the near future more direct methods for retarding turnover could be introduced (Reeds, 1987).

Disease

Although great progress has been made in the control and eradication of cattle diseases, improvements have been uneven (e.g. eradication only in isolated countries), expensive (vaccination programmes) and impermanent. The greatest scourge of cattle, foot and mouth disease (FMD), illustrates the difficulties. The most fortunate countries have either prevented the entry of FMD or succeeded in eradicating it by vaccination and/or compulsory slaughter, but those with no natural barriers to re-infection must continue to vaccinate against FMD. Many countries will never eradicate it because it is endemic in wild animals (compare to smallpox in man). For FMD and similar diseases (e.g. rinderpest) the best one can hope for in the future is that cheaper and more effective vaccines will further restrict the areas affected.

As some diseases are suppressed or eradicated others assume greater importance. In the developed countries the so-called 'production diseases' have become prominent as production per animal has increased. Also, new diseases may appear, the latest example for British cattle being bovine spongiform encephalopathy. Increasing concern for animal welfare puts greater emphasis on controlling disease to give animals a better life (not just to prevent economic loss or the transfer of disease to man). Many of the cattle diseases now rising to the top of the rank order (e.g. mastitis) are multifactorial and cannot be controlled through the imposition on producers of a national prevention programme. The onus for removing the constraint of such diseases is placed on the individual producer. Ideally, diseases are prevented rather than treated, but preventive medicine schemes for cattle have been adopted only at a slow rate, even by the large-scale, educated producers of northern Europe. Larger herds and fewer stockmen make the detection of disease more difficult. If fully automated milking is introduced, parlours will need to include automated devices for detecting mastitis.

Disease constraints on animal production will therefore remain, especially in developing countries. In those countries a disproportionate amount of the money available for developing animal production must often be used to control disease (Crotty, 1980). New weapons for the fight against disease should come from genetic engineering; they include the vaccines mentioned above, diagnostic aids, and anti- or pro-biotics. In addition, it may prove possible in the future to make more progress in the breeding of disease-resistant animals. We already recognize and value the ability of some types of cattle to tolerate trypanosomiasis and resist tick infestation. The latter characteristic is highly heritable, but selection for resistance to diseases like mastitis or bloat is difficult because the character is multigenic and of low heritability. It may be argued that selection for production traits such as milk yield will carry with it some associated selection for disease resistance. However, if this is the case, newer and more specific procedures for increasing production traits – such as the transfer of the specific genes involved – will need to be accompanied by more deliberate efforts to promote disease resistance.

Conclusions

World cattle production systems still fall into two major groups. The first group includes most of the systems used in the developed countries, which are geared to high rates of production and dependent on expensive inputs. The expansion of these systems is not limited by the availability of resources or techniques, but by the restricted market for their high-cost products. However, economic constraints are not sufficiently severe to drive production systems into a single mould. In the future, the cattle production systems of the developed countries will continue to be at the 'sharp end' of technology, and seem likely to benefit from new techniques, especially in the areas of genetic improvement and reproduction. But producers will need to adapt their systems to the social constraints arising from consumers' concern for the health and welfare of farm animals and of themselves. In the longer term man may cease to eat beef.

The second group consists of most of the cattle production systems of the developing countries, which are restricted to low productivity by a variety of constraints. One can argue that these systems need little improvement because they are already efficient in utilizing resources that would otherwise be wasted, and they provide essential nutrients and draught power for their owners. Yet because of human population growth these systems must improve merely to maintain their contribution to man's needs, and must improve still further if they are to promote the economic advancement of developing countries. Moreover, improvements for these systems must be in production per animal, not in numbers of animals (as over-stocking is already a chronic problem).

Many of the constraints on cattle production systems in the developing countries are either structural (imposed by social customs and by national and international economic policy) or due to lack of resources, and cannot be removed by means of new techniques. Preston and Leng (1986) show how old and new technology could be used to make the best use of the existing feed resources of the developing countries. Yet without new resources (e.g. concentrate feeds) it is probably unrealistic to expect any major 'leap forward' in production per animal, for even in the developed countries (with their better quality forages) it is unusual to find efficient and profitable systems based almost entirely on these forages. New techniques in animal breeding, including genetic engineering, are unlikely to have much impact on less intensive cattle systems unless they are strikingly radical but simple (e.g. involving rumen fermentation of fibre). Even new disease control techniques will be ineffective unless supported by improvements in nutrition.

The structural constraints overshadow the technical problems of cattle production in developing countries. International trade in cattle products is severely restricted; thus even Australia – with vast resources that include cheap cereals – is unable to develop its cattle industry to its full potential. In many countries, cattle production is caught in the vicious circle through which lack of agricultural development fails to free people to work in manufacturing industry, which in turn prevents the growth of consumer demand for beef and milk. In countries with a predominantly pastoral agriculture, the responsibility for breaking out of this circle may lie heavily on the cattle industry, yet from past performance it seems unlikely to be successful in achieving the required improvements in productivity.

References

Crotty, R. (1980) *Cattle, Economics and Development.* Farnham Royal: Commonwealth Agricultural Bureaux

De Boer, F. (1987) Perspectives of bovine production: from cow to supercow? In *Future Production and Productivity in Livestock Farming: Science versus Politics.* Elsevier, Amsterdam, pp. 21–36

Edinburgh School of Agriculture (1988) *Langhill Past, Present and Future.* Mimeographed Report, Edinburgh School of Agriculture

Food and Agriculture Organisation (FAO) (1964; 1965; 1985) *Production Yearbook.* FAO, Rome

Gregg, K., Bauchop, T., Hudman, J. F., Vercoe, P. *et al.* (1987) Application of recombinant DNA methods to rumen bacteria. In *Recent Advances in Animal Nutrition in Australia 1987*, (ed. D. J. Farrell), Armidale, University of New England

Holmes, J. H. G (1983) Animal production and the World food situation. In *World Animal Science (Vol. A1): Domestication, Conservation and Use of Animal Resources*, (ed. L. Peel and D. E. Tribe), Elsevier, Amsterdam, pp. 255–284

Jasiorowski, H. A. and Quick, A. J. (1987) Cattle production systems in practice. In *World Animal Science (Vol. C3) Dairy Cattle Production*, (ed. H. O. Gravert), Elsevier, Amsterdam, pp. 269–289

Milk Marketing Board (1985) *An Analysis of Farm Management Services Costed Dairy Farms 1984–5.* Farm Management Services Information Unit, Reading, UK

Ministry of Agriculture, Fisheries and Food (MAFF) (1987) *Household Food Consumption and Expenditure 1985.* Annual Report of the National Food Survey Committee. HMSO, London

New Zealand Dairy Board (1985) 61st Livestock Improvement Report, Hamilton, New Zealand

Nicholas, F. W. and Smith, C. (1983) Increased rates of genetic change in dairy cattle by embryo transfer and splitting. *Animal Production, 36*, 341–353

Paul, A. A. and Southgate, D. A. T. (1978) *McCance and Widdowson's The Composition of Foods.* HMSO, London

Preston, T. R. and Leng, R. A. (1986) *Matching Livestock Production Systems to Available Resources.* International Livestock Centre for Africa, Addis Ababa, Ethiopia

Reeds, P. J. (1987) Metabolic control and future opportunities for growth regulation. *Animal Production, 45*, 149–169

Taylor, St. C. S., Moore, A. J., Thiessen, R. B. and Bailey, C. M. (1985) Efficiency of food utilization in traditional and sex-controlled systems of beef production. *Animal Production, 40*, 401–440

Woolliams, J. A. and Smith, C. (1988) The value of indicator traits in the genetic improvement of dairy cattle. *Animal Production, 46*, 333–345

World Bank (1986) *World Development Report 1986.* University Press, Oxford

Chapter 2

Dual purpose cattle production systems

T. R. Preston and Lucia Vaccaro

In designing production systems for cattle the changing role of livestock, in both industrialized and Third World countries, must be considered. Food production is no longer the first priority for the majority of countries; and even in those regions where famine appears to be endemic, attention must be focused on the overall issues of rural development and preservation of the environment, and not simply on how to increase supplies of milk and meat.

In many Third World societies, cattle are more important as a source of manure – for fuel and/or fertilizer – and power, than of milk and meat. For the rural poor, they are more secure than the bank, as a means of safeguarding savings from inflation and devaluation.

The industrialized countries, grappling with over-production of almost all agricultural and livestock products, and with increasingly vociferous lobbies in favour of animal and environmental welfare, and the need to increase employment opportunities in rural areas, would also do well to reflect on the possibilities of cattle production systems playing a slightly different role than that of specialist milk and meat producers.

It will be argued that it is more economical, in terms of national resource utilization, and with the above constraints in mind, to satisfy the demand for milk and beef by combining both activities in the same animal. The justification for this approach is that:

(1) the target levels of production – 2000 litres of milk and 300 kg of beef/cow per year – are closely related to national demand ratios which vary from 4 to 5 litres milk/1 kg of beef
(2) advantage can be taken of important physiological traits, previously disregarded in intensive specialized systems – e.g. the effect of suckling in stimulating milk yield, reducing stress in both cows and calves and permitting the calf to use more efficiently supplementary feed of low protein content.

Of special importance to Third World countries is that breeding programmes for dual purpose milk–beef systems permit a much greater degree of self-reliance (i.e. reduced dependence on expensive (imported) inputs), and the technology is simpler – and therefore more easily applied and with greater chance of acceptance – than for specialized systems, especially milk production.

Technology transfer and training

An international review of new techniques in cattle production is expected to have an impact in two major areas:

transfer of technologies on farms
training of professionals in animal production.

Because it is international, the target audience must be defined and the messages interpreted for and directed to the respective target groups. These can be classified broadly as:

industrialized countries
developing countries.

It would be convenient, and reassuring also, to believe that both the goals for, and the means of delivering, such technologies would be the same for both industrialized and developing countries. Unfortunately, experience shows quite clearly that both the objectives and the means differ according to the target group of countries. Even if it were desirable to achieve the standard of living presently enjoyed by most industrialized countries – and opinions differ widely on this issue – there are not the resources (meaning mainly fossil fuel) nor the wealth with which to attain and sustain such levels of consumption on a world basis.

The major part of the population of most developing countries still lives in rural areas. Livestock are an integral part of rural living. They may be the main source of draught power for agriculture; they are at the same time a bank and a source of credit; they provide fuel; they permit the upgrading of non-edible (by humans) feeds into high quality nutrients which are indispensible for children and pregnant women. They help a poor family to generate income with which to buy food in order to subsist.

In these roles, livestock are vitally important. However, in their impact on the overall food supply for a community, their presence at times can be extremely negative. Overgrazing of community land by cattle and small ruminants is one of the major causes of erosion and other ecological damage in Africa, which in turn directly threatens food supplies. Similar negative circumstances are arising in much of tropical Latin America and Australia, where cattle ranches have replaced forests.

Our recommendations in this review as to what constitute new cattle techniques may therefore have far-reaching consequences. As contributors we have major responsibilities because of:

the direct impact of a new technology through its immediate transfer to a developing country
indirect effects through teaching and training.

There is cause for concern in both areas (Preston and Leng, 1987; Timberlake, 1988). Experience has proved that for those developing countries situated in the tropics, the livestock production models developed in the industrialized countries have mostly proved inappropriate, leading to ever increasing dependency rather than the greater self reliance which is sought. Moreover, this fundamental problem is exacerbated by the inappropriate training of professional people, especially those in the animal sciences.

Why is this so? And what can be done about it?

In Table 2.1 some of the major differences between industrialized and developing countries are summarized; these make it difficult to transfer technologies derived in the former. These differences help to explain why there are conflicts concerning the strategies that should be applied when the aims are to establish appropriate procedures in research, extension and training on behalf of the developing countries.

Table 2.1 Livestock production in industrialized and developing countries; goals and means

	Industrialized	Developing
Climate	Mostly temperate	Mostly tropical
Role of livestock	Specialized	Multipurpose
Target group	The rich	The poor
Resource base:		
feed	Starch–protein	Fibre–sugars
genetic	Improved	Native
Capital intensity	High	Low
Labour intensity	Low	High
Mechanization	High	Low
Agrochemicals	High	Low
Infrastructure	Good	Poor
Marketing	Good	Poor

Of particular relevance to this book is the more varied and often quite different role of livestock, especially cattle, in developing as opposed to industrialized countries. In the former, they mostly favour the poor; in the latter the affluent, who have the purchasing power to afford what are fast becoming luxuries in the human diet.

The feed resource base is also vastly different. Starch- and protein-rich feeds, which are the major components of diets in industrialized countries, are difficult to grow in the tropics and are also the staples of the human diet. Therefore the cost of such feeds in those tropical countries where they are not subsidized (usually by importing them), is disproportionately high in relation to the cost of the locally available biomass, in the form of pastures and forages. For example, in Colombia the opportunity price of sorghum grain is US$250/tonne and of soyabean meal US$450/tonne. Sugar cane by contrast is worth US$40/tonne (dry matter basis) delivered to the factory. Rice straw, harvested, baled and transported to a central site on the farm will cost US$30/tonne.

Other differences relate to the use of capital and labour. In the industrialized countries, inputs of the former are high and of the latter low; exactly the opposite holds for most of the developing countries. Commercial interest rates in Colombia are 4%/month and the theoretical minimum wage (including social benefits which rarely are paid in rural areas), is less than US$4.00/day. Machinery (mostly important or assembled from imported components) is often heavily taxed, to conserve foreign exchange. The same is true of drugs and other agrochemicals, including fertilizers.

These are some of the reasons why a different strategy is needed.

Towards a more appropriate livestock development strategy in tropical developing countries

The first steps to take are as follows:

identify the potential and available local (on-farm) resources
match the production system with the resources available.

In the tropics, this latter approach will generally mean the development of integrated and mixed farming systems, which permit more efficient use of the basic resources of solar energy, rainfall and soil. The typical feed resources available in the tropics, are summarized in Table 2.2. None of these resources is suitable for high level milk production. In all of them, the array of available substrates is imbalanced at the level both of the rumen (for fermentable nitrogen and micronutrients) and the animal (bypass protein, glucogenic precursors and long chain fatty acids (Preston and Leng, 1987).

Table 2.2 Typical feed resources available in the tropics (medium to high rainfall (>500 mm) and irrigated areas)

Forage crops, especially sugar cane
Crop residues
Agroindustrial byproducts
(Pasture)

Finally, the feed supplements which would be needed to correct the imbalances are much too expensive to permit their use at other than strategic minimal levels; i.e. the point of economic equilibrium is reached at a level of productivity much below the biological maximum.

In most tropical Third World countries, it is therefore pointless to discuss raising potential yields by 20–30%, either by genetic improvement or direct hormone manipulation, when the present ceiling (assumed to be of the order of 5000 kg/lactation) already exceeds the limits set by nutrition, climate and management.

The biological arguments are stacked heavily against specialist dairying in the tropics. However, there are equally strong economic and sociological reasons which favour dual purpose systems. A good point at which to begin is to examine the demand ratios for milk and beef in selected countries (Table 2.3).

Table 2.3 Relative per caput demand for milk and beef in one developed and two developing countries

	USA	Colombia	Pakistan
Demand (kg/caput per year) for:			
milk	250	70	54
beef	53	18	14
Milk:beef demand ratio	4.7	3.9	3.9

Source: FAO (1980).

It appears that in spite of wide differences in purchasing power and cultural preferences, the consumption pattern for these two commodities is relatively constant at approximately 4 kg milk/1 kg of carcass beef. In specialized milking herds, averaging 5000 kg/cow per year, assuming the male calf is raised for meat, the production ratio of milk to beef is approximately 20:1, and even wider if the male calf is slaughtered at birth – which is a common practice in many specialist dairy enterprises in developing countries, especially when the breeds used are extreme dairy type, i.e. North American Holstein and Jersey. The result is that in order to balance the milk:beef supply to meet demand, then some four specialist beef cows must be kept for each dairy animal. In the industrialized countries the net result of such strategies has been a decline in dairy cattle numbers and an increase in the beef herd (Table 2.4).

Table 2.4 Changes in the populations of dairy and beef cattle in the USA

Year	Dairy cows (millions)	Beef cows (millions)
1957	35	60
1960	32	65
1965	27	81
1970	22	90

Source: FAO (1980).

Apart from the biological and economic inefficiency of specialist beef production, such systems are especially undesirable in developing countries since it is invariably the 'ranchers' and 'latifundistas' who most oppose land reform; at the same time, it is the extensive grazing systems which have been the stimulus for much of the deforestation, and subsequent erosion, in vast areas of the tropics.

It may be claimed that the new technologies, discussed in this book, will lead to dramatic improvements in the biological efficiency of beef cattle. But again, a note of warning must be sounded, for all the evidence so far suggests that it is the rich 'hobby' farmers who benefit most from the capital-intensive procedures that are the basis of these 'new technologies'.

Dual purpose cattle production systems in the tropics

Dual purpose cattle production systems are those in which income is divided approximately equally between milk and beef. They are predominant in many parts of Latin America (Table 2.5). Their salient characteristics are that almost invariably the calves are raised on the cow by some form of restricted suckling. Usually milking is only once daily and the major feed resources are pasture or fibre-rich crop residues and byproducts with minimum use of supplements.

The genetic resources vary enormously, but the most popular animals for this system are crossbreds, derived from European *Bos taurus* types (Brown Swiss and Holstein predominantly) and *Bos indicus* (Zebu). Typical performance data from a number of countries are summarized in Table 2.6.

Table 2.5 Cattle production systems in the coffee-growing region of Colombia

	Altitude (m above sea level)			
	>2000	1250–2000	<1250	Total
Proportion (%) of farms dedicated to:				
Specialized beef	8	2	10	20
Specialized milk	9	9	2	20
Dual purpose	25	22	13	60

Source: Suárez and Jaramillo (1988).

Table 2.6 Typical performance data for cattle managed according to the dual purpose system on demonstration or experimental farms in a number of tropical countries

	Milk/year		Weaning weight (kg)	Calving interval (days)
	Saleable (kg)	Calf (kg)		
Dominican Republic[1]	1750	470	165	380
Mexico[2]	1400	450	150	401*
Costa Rica[3]	1300	400	155	400*
Malaysia[4]	1860	?	?	438*

* Exclusively with artificial insemination (AI).

[1] Fernandez, Macleod and Preston (1977).
[2] Alvarez et al., (1980).
[3] M. E. Ruiz, personal communication.
[4] Cheah and Kumar (1984).

The dual purpose system arose through the need to increase the income from typically extensive beef production systems. Often the first stage is the milking of a proportion of the cows – those with appropriate genetic potential and temperament are chosen for this purpose. The next step is usually to introduce a sire from a recognized dairy breed, in order to increase dairy traits. Further innovations may follow, such as pasture improvement, supplementation of cows and calves, twice daily milking and occasionally machine milking.

More recently (Preston, 1977), dual purpose systems have been advocated as an appropriate way to integrate cattle into intensive mixed farms, especially in the wet tropics. The arguments used are that such systems enable better use to be made of available resources, that they are well understood by farmers (who developed them in the first place) and that they satisfy the demand ratio for milk and beef, as described earlier.

Aside from these economic considerations, there are distinct biological advantages intrinsic to dual purpose systems. These features are not well known and even less well understood. It is important to describe them, so that those scientists who are in research centres in industrialized countries, who have the necessary laboratory resources and expertise, may feel stimulated to direct some of their attention to these areas with a view to establishing the underlying mechanisms.

Restricted suckling

Effects on the cow

Use of the calf to stimulate milk letdown is the traditional technique employed to coax beef animals to surrender a part of their milk output for human consumption. In crossbred cattle derived from Zebu (*Bos indicus*), typically used in dual purpose systems, there appears to be a negative linear relationship between the proportion of genes derived from the *Bos taurus* parent and the incidence of short lactations (Table 2.7).

Table 2.7 Effect of genetic make-up on incidence of short lactations in Holstein/Zebu crosses in Mexico

Proportion of Holstein genes (%)	Incidence of short (>70 days) lactations (%)
25	76
50	40
75	10
100	None

Source: Alvarez *et al.* (1980).

Table 2.8 Milk production from F1 European (Holstein or Brown Swiss)/Zebu crosses milked with and without calf stimulation

	33 first-calvers	
First lactation:	without calf-stimulation	
	16 milked adequately	17 became dry <100 days
Second lactation:	Without calf stimulation	With calf stimulation
Prematurely dry <100 days	8	0
Lactation length* (days)	216	270
Total milk* (kg)	590	1680
Saleable milk* (kg)	590	1000

* For the cows which milked more than 100 days.

Source: Alvarez *et al.* (1980).

In an unselected F1 herd (derived by crossing Zebu females with Holstein and Brown Swiss sires), milked by machine (Table 2.8), half the animals had lactations lasting less than 70 days when the calf was not present at milking. In their second lactation those cows, which had had short lactations previously, milked normally when the calf was used to stimulate letdown. By contrast, the cows which milked normally in their first lactation (without calf stimulation), regressed to the mean in their second lactation, half of them becoming dry before 70 days.

As well as ensuring normal length lactations in crossbred cattle, restricted calf suckling brings other benefits. In a recognized dairy breed (e.g. Holstein), cows that suckled their calves after milking gave more milk during the period that suckling was practised and subsequently after the calf had been weaned (Table 2.9). There is less mastitis in cows that are milked and also suckle their own calves or calves from other cows (Table 2.10) compared with cows that are milked by hand or machine but do not suckle.

Table 2.9 Effect of two systems of restricted suckling on milk yield of Holstein cows and milk intake by their calves

	Control (did not suckle)	Suckled	
		Twice daily for 70 days	Twice daily for 28 days then once daily for 42 days
Saleable milk (kg/day)			
5–28 days	12.5	9.7	9.5
29–70 days	11.5	9.5	13.5
71–112 days	10.0	11.8	12.9
Consumed by calf (kg/day)			
5–28 days	–	5.8	5.4
29–70 days	–	6.3	2.5
Total milk yield (kg/day)			
5–28 days	12.5	15.5	14.9
29–70 days	11.5	15.8	16.0
71–112 days	10.0	11.8	12.9

Source: J. Ugarte and T. R. Preston, unpublished data.

Table 2.10 Effect of suckling on incidence of subclinical mastitis (expressed as % of all quarters examined) in F1 (European × Zebu) and Holstein cows in the tropics

Authors	Calf suckling		Breed
	No	Yes	
Alvarez *et al.* (1980)	21	6	F1 (E × Z)
Ugarte and Preston (1972)	6	2	Holstein
Ugarte and Preston (1975)	8	2	Holstein

If cows which suckle their calves give more milk than those which do not suckle, then it would be expected that either they must eat more food or mobilize more body tissue. However, in an experiment designed to test this hypothesis (Table 2.11), Holstein cows that suckled their calves after machine milking, gave more milk and lost less weight immediately after calving than cows which had their calves removed permanently 3–5 days after birth. The differences in bodyweight continued to be manifested at least through the first 3 months of lactation. Feed intake was maintained constant in both groups. The implication is that the stress on

Table 2.11 Effect of suckling on milk production and bodyweight change in Holstein cows in Venezuela. (The control cows had their calves removed permanently after the first 4 days; the experimental group suckled their own calves for 20 minute periods twice daily immediately after the cows had been machine-milked)

	Control (no suckling)	Restricted suckling	SEM
Milk production (kg/day)			
saleable	7.9	9.0	±0.8
consumed by calf	4.0	6.1	
Total	11.9	15.1	
Liveweight change (kg)			
Pre- to 7 days post-partum	−72	−46	±15
From 7 to 84 days post-partum	+15	+3	±5

Source: Velazco, Capriles and Preston (1982).

the dam caused by taking away its offspring led to adrenalin-stimulated demand for glucose, resulting in increased mobilization of body reserves.

Effects on the calf

Efficiency of milk utilization is higher in calves that are suckled than when they take the same amount of milk from a bucket (Table 2.12). This is understandable in the light of Ørskov's work (Ørskov, 1983) which demonstrated that psychological stimuli, rather than physical factors, were the mechanisms which controlled the closing of the oesophageal groove which directs milk to the abomasum. Bucket feeding, by contrast, results in much milk spilling over into the rumen where the fermentative mode of digestion leads to losses in both the quality and quantity of nutrients available to the animal.

Other benefits are a reduced incidence of diarrhoea and elimination of navel sucking, as a result of which suckled calves can be housed in groups, permitting lower investment in housing, simpler feeding and management and less stress on the calves.

Table 2.12 Calves use milk more efficiently by suckling rather than by bucket feeding (calves were crossbred European × Zebu raised from birth to 84 days of age either by bucket feeding of whole milk or by restricted suckling for 20 minutes following milking)

	Bucket	Suckling	SEM
Condition score*	1.61	1.35	±0.04
Milk intake (kg/day)	3.08	2.73	±0.12
Milk conversion (kg milk/kg liveweight gain)	9.7	4.9	±1.0

* Belly girth (cm)/liveweight (kg): low value = more tissue and less gut fill.

Source: Fatullah Khan and T. R. Preston, unpublished data.

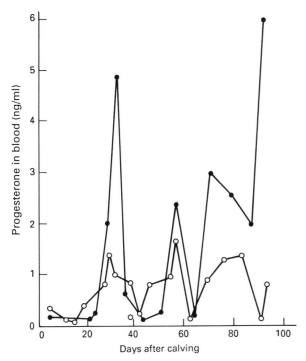

Figure 2.1 Effect of restricted suckling on blood progesterone levels. ○ Cows that suckled their calves after milking; ● control cows that had their calves separated 4 days after parturition. (Velazco *et al.*, 1982)

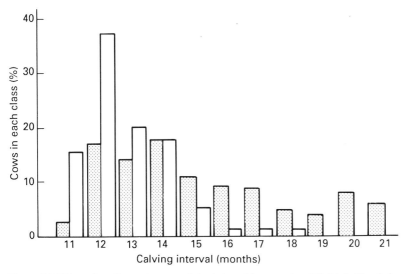

Figure 2.2 Effect of mating system on calving interval in a crossbred Holstein/Creole herd managed with restricted suckling according to the dual purpose system. ▦ Artificial insemination; ☐ natural mating. (Naidoo, Hulman and Preston, 1981)

Disadvantages of restricted suckling

Poorer fertility is generally ascribed to calf suckling due to extension of the interval between calving and conception. It is generally believed that for intensive cattle production systems in industrialized countries this is due to delay in initiation of ovarian activity. However, there is some evidence that the impaired fertility with calf suckling in developing countries is due not so much to delay in ovarian activity, but to poor manifestation of oestrus (silent heat) due to a reduced amplitude of the progesterone peaks which regulate ovarian cycles (Figure 2.1).

Use of natural mating rather than artificial insemination is therefore advocated in dual purpose systems. This is substantiated by observations in a dairy enterprise in Mauritius where calves were raised by restricted suckling (Figure 2.2). When there was exclusive use of artificial insemination (AI), calving intervals were long and variable; running bulls with the herd reduced both the average calving interval and variability.

Restricted suckling in *Bos taurus* herds

Modifications to the management of cows and calves may be needed when calf suckling is introduced into herds in which the cows are mainly of *Bos taurus* origin and therefore do not need the physical presence of their calves to stimulate milk letdown. In such cases the calves are suckled when milking is completed, either in the shed where the cows are milked or in a pen designated for that purpose. It has been observed that in this system, up to 20% of cows may withhold most of their milk during milking, retaining it for their calves. This problem can be overcome by cross-suckling, in a way which does not allow cows to suckle their own offspring (E. Murgueitio, unpublished data). For example, the cows in early lactation suckle the calves from cows in late lactation, and vice versa.

Breeding programmes for dual purpose systems

There is now broad agreement that in the humid tropics, the most appropriate animals for dual purpose systems are those derived by crossing native cattle (usually *Bos indicus*) with any of the recognized dairy breeds; and that the optimum proportion of European genes will vary according to the harshness of the environment (McDowell, 1985). The results from the 'on-farm' evaluations in Brazil, made by Madalena *et al.* (1982), show that there are few advantages and many disadvantages when the proportion of genes from a specialized European dairy breed exceeds 50%.

The most popular crossing sires are Holstein, Brown Swiss, Normandy and Simmental. Few breed comparisons have been made but the more reliable data indicate a significant advantage to the use of Holstein sires compared with Brown Swiss (Vaccaro, 1984). While there are many who advocate the merits of the native Criollo breeds in Latin America, their numbers are small and there are almost no data which permit valid comparisons to be made with other breeds and crosses (Vaccaro, 1987).

It is frequently argued that it is difficult to stabilize a cattle population in order to maintain approximately equal proportions of *Bos taurus* and *Bos indicus* genes. However, in practice this is not a major problem, once it is accepted that the appropriate way is by using F1 bulls. The recommended system is to 'manufacture' such bulls by crossing native 'adapted' females with imported semen from progeny tested sires of the selected European breed (L. Vaccaro, personal communication). Hardiness and fertility are ensured by selecting the female parent for these characteristics. A sustainable level of milk production (1000–1500 kg/lactation) is guaranteed by choosing semen from a bull with a proven capability to maintain yields (in pure bred dairy females) of about 5000 kg/lactation (the potential yield of the F1 offspring is then at least 2500 kg, ignoring both the dam's contribution and the effects of heterosis). Almost all dairy bulls presently standing at approved insemination centres in the industrialized countries have this capacity. F1 bulls can be run with the herd which facilitates natural mating. This is recommended in view of the difficulties of heat detection in cattle that raise their calves by restricted suckling systems.

Conclusions

Dual purpose systems for combined milk and beef production offer many advantages to tropical Third World countries, since they permit a much greater degree of self-reliance (i.e. reduced dependence on expensive (imported) inputs), and the technology is simpler – and therefore more easily applied and with greater chance of acceptance – than for specialized systems, especially milk production. The basic justification for the dual purpose concept is that the target levels of production – 2000 litres milk and 300 kg beef/cow per year – are closely related to national demand ratios which vary from 4 to 5 litres milk/1 kg of beef. The total cattle population required to support these yield levels is no higher than if the milk and beef were produced in separate herds, but with the additional benefits (for most Third World countries) of supporting more employment opportunities and enabling greater and more efficient use to be made of presently under-utilized local feed resources. Another important issue is that advantage can be taken of important physiological traits, which have been disregarded in intensive specialized systems, e.g. the effect of suckling in stimulating milk yield, reducing stress in both cows and calves and permitting the calf to use more efficiently supplementary feed of low protein content.

Breeding programmes for dual purpose milk–beef systems are simple and low cost because they take advantage of F1 sires produced by combining imported 'proven' (for milk!) semen with the adaptability and fertility of native females. This avoids the need to set up national progeny testing schemes which besides being expensive are also unreliable due to the difficulty of obtaining the necessary herd records.

References

Alvarez, F. J., Saucedo, G., Arriaga, A. and Preston, T. R. (1980) Effect on milk production and calf performance of milking crossbred European/Zebu cattle in the absence or presence of the calf, and of rearing their calves artificially. *Tropical Animal Production*, **5**, 25–37

Cheah, P. F. and Kumar, R. A. (1984) Preliminary observations on the performance of Sahiwal × Bos taurus dairy cattle. *Malaysian Veterinary Journal*, **16**, 1–7

Food and Agriculture Organization (FAO) (1980) Production yearbook. FAO: Rome

Fernandez, A., MacLeod, N. A. and Preston, T. R. (1977) Production coefficients in a dual purpose herd managed for milk and weaned calf production. *Tropical Animal Production*, **2**, 44–48

McDowell, R. E. (1985) Crossbreeding in tropical areas with emphasis on milk, health and fitness. *Journal of Dairy Science*, **68**, 2418–2435

Madalena, F. E., Valente, J., Teodoro, R. L. and Monteiro, J. B. N. (1982) Milk yield and calving interval of Holstein-Friesian and crossbred Holstein-Friesian:Gir cows in a high management level. *Pesquisa Agropecuaria Brasileiro*, **18**, 195–200

Naidoo, G., Hulman, B. and Preston, T. R. (1981) Effect of artificial insemination or natural mating on calving interval in a dual purpose herd. *Tropical Animal Production*, **6**, 188

Ørskov, E. R. (1983) The oesophageal groove reflex and its practical implications in the nutrition of young ruminants. In *Maximum Livestock Production from Minimum Land* (ed. C. H. Davis, T. R. Preston, M. Haque and M. Saadullah), Bangladesh Agricultural University, Mymensingh, pp. 47–53

Preston, T. R. (1977) A strategy for cattle production in the tropics. *World Animal Review*, **21**, 11–17

Preston, T. R. and Leng, R. A. (1987) *Matching Ruminant Production Systems with Available Resources in the Tropics and Sub-tropics*. Penambul Books, Armidale, Australia, p. 245

Suarez, S. and Jaramillo, C. J. (1988) Algunas caracteristicasde la explotacion ganadera en la zona cafetera de Colombia. *Pasturas tropicales*, **10**, 24–27

Timberlake, L. (1988) Africa finds a new vicious circle. *New Scientist*, 4 August, p. 65

Ugarte, J. and Preston, T. R. (1972) Rearing dairy calves by restricted suckling once or twice daily on milk production and calf growth. *Cuban Journal of Agricultural Science*, **6**, 173–182

Ugarte, J. and Preston, T. R. (1975) Restricted suckling VI: effects on milk production, reproductive performance and incidence of clinical mastitis throughout the lactation. *Cuban Journal of Agricultural Science*, **9**, 15–26

Vaccaro, L. (1984) El comportamiento de la raza Holstein Friesian comparada con la Pardo Suiza en cruzamiento con razas nativas en el tropico: una revision de la literatura. *Produccion Animal Tropical*, **9**, 93–101

Vaccaro, L. (1987) Aspectos geneticos del programa de investigacion en bovinos de doble proposito en el ICA. *Consultancy Report*. ICA: Bogota, 26 pp

Velazco, J. A., Capriles, M. and Preston, T. R. (1982) Efecto del amamantamiento restringido sobre la produccion de leche y cambio de peso vivo en vacas Holstein y Pardo. *Jornadas Tecnicas Instituto de Produccion Animal*. UCV, Maracay

Velazco, J., Calderon, J., Guevera, S., Capriles, M. *et al.* (1982) Efecto del amamantamiento restringido sobre el crecimiento de becerros, produccion de leche, actividad ovarica y nivel de progesterona en vacas Holstein y Pardo Suizo. In *Informe Anual Instituto de Produccion Animal*. UCV, Maracay, pp. 66–165

Chapter 3

New techniques in the mechanization and automation of cattle production systems

S. L. Spahr

The application of automation and mechanization to cattle production systems is accepted by many as the single most important factor affecting cattle production practices since the mid-1970s. The effect is most dramatic in the management of dairy cattle, with similar trends reported from many countries. In the USA in 1985 labour per cow was 42% and labour per unit of milk only 33% of the amount used in 1975. During this same period, average herd size increased from 78.5 to 88.2 and milk production per cow increased from 6183 kg to 7383 kg in herds on the national milk records system (DHI) test (about 4.5 million cows).

The increases in labour efficiency and accompanying reductions in the drudgery required for routine care of cattle have gradually changed the role of the manager for cattle production units, resulting in a higher level of management skills being required today than a decade ago. Advanced production techniques built around

Table 3.1 Automatic equipment for improving the management of cattle production

Production function	Equipment
Feeding	Automatic concentrate dispensers
	Electronic data acquisition of amount of feed dispensed per location on mobile feed mixer-dispensers
	Automatic milk replacer dispensers
Range or free roaming animal monitoring	Electronic animal identification
	Automatic animal weighing and daily communication to remote management computer
	Activity counters (oestrus detection)
	Implanted impedance sensors (oestrus detection)
Milking parlour	Automatic detachers
	Power gates
	Electronic milk meters
	Backflush units
	Evaporative coolers
	Udder health sensors
	Robot milker

mechanization and automation have been developed and adopted on a widescale for feeding, milking, and data analysis (Table 3.1). New methods for electronic monitoring of the reproduction status of cows and heifers, for more precise selection of sires based on trait-by-trait computerized selection criteria, and for national systems of identifying cattle via electronic identification are being actively pursued.

Feeding and housing

The trend towards mechanization and automation in cattle production systems has been fuelled by a desire on the part of the herd manager to reduce labour and drudgery and to improve animal performance and animal comfort. These goals have manifested themselves in the form of more confinement housing and mechanized feed delivery systems (Figure 3.1).

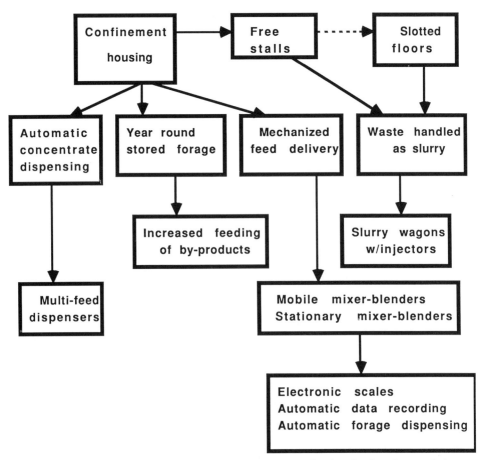

Figure 3.1 Housing and materials handling technologies accompanying the adoption of new techniques in cattle production

Early equipment for mechanized feed delivery was built around electrically powered augers for feeding of groups of animals. Recent systems for group feeding may be categorized into one of two groups:

(1) stationary systems which feature stationary mixer-blender units with electronic scales to provide precise weighing of feed ingredients. Belt systems have now replaced augers as the method of choice for feed delivery and many of the on-farm systems have automatic features installed to minimize operator involvement in the mechanics of moving individual feeds to the mixer, accurately weighing them, and delivering the mixed feed (a total mixed ration) to the appropriate location (Puckett, Spahr and McCoy, 1987)

(2) mobile mixer-blender units have also become a standard piece of equipment in many cattle housing units. Most mobile mixer units are also equipped with electronic scales for precise and accurate weighing of feed ingredients. This system ranges from the small mixer-dispensing unit pulled by tractors to the large self-propelled units found in Israel which are equipped with an on-board computer to communicate the mixture and the amount to be fed for each group to the operator, and to provide an automatic recording of the amount actually dispensed at each location (Feedtrol, Givot Brenner, Israel). One effect of these changes is that the feeding of byproduct feeds (from wet-corn processing, and low-quality local produce) has been made much easier and more flexible. Totally automated systems for feeding stored forage have been developed (Forbes *et al.*, 1986; Ipema and Rossing, 1987), but are still at the research station level (Figure 3.2).

Figure 3.2 On board computer with RS 232 communication interface for recording amounts of various feeds dispensed to each of several feeding locations. Reports may be prepared for amounts of each feed actually fed relative to the amount programmed. Amounts programmed may be entered electronically from a personal computer. (Courtesy of Gavich Software and Equipment Development)

Automatic individual-animal feed dispensers have been developed and are widely accepted in most developed countries (Smith and Pritchard, 1983). These systems depend on automatic individual identification of each animal and are controlled either by standard desktop computer or by proprietary processors. In their simplest form these units dispense a concentrate mixture to each cow on an automatic individual basis according to allocations entered on the processor by the herd manager. In the recent models more than one feed ingredient (up to four in some models) may be dispensed simultaneously allowing each cow to have a separate concentrate mixture tailored to her individual nutrient needs for energy, protein and minerals. This system is particularly appealing in situations where it is desired to add specific costly ingredients (e.g. rumen bypass protein, rumen bypass fat or specific additives) to the diet of high producing cows, but not to low producers. High moisture grains also may be fed automatically, individually through some units.

Automatic individual concentrate dispensing has found its greatest popularity in dairy herds of approximately 60–150 cows which are housed in groups. Herds smaller than 60 usually either use individual stall housing (stanchion barns) or cannot afford the investment required for automatic individual concentrate dispensing. Herds larger than about 150 cows usually can achieve satisfactory feeding programs for various daily milk production levels more economically by grouping their cows according to nutrient need and feeding total mixed rations via mobile feed mixer-dispensing units.

Feeding behaviour with automatic concentrate dispensing is affected by the control program for the specific unit. Typical features of the control program include dividing the concentrate into several small meals (six meals/day is a typical situation, but many systems have the capability for up to 24), automatic recording of the amount of feed dispensed for each cow, and a daily summary of cows which did not receive their entire allotment. In our experience, cows typically enter the dispensing stalls 12–15 times daily. However, individual animal usage varies widely. Some cows enter the stalls 35–40 times daily while others (particularly

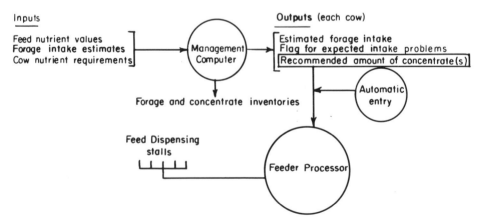

Figure 3.3 Example of 'closed loop' ration balancing for automatic individual cow concentrate dispensing. Concentrate recommendations are derived electronically using nutrient requirements of each animal, nutrient composition of the feeds and estimated feed intakes of each animal. Concentrate recommendations are transferred electronically to the concentrate dispenser system to eliminate manual entry

those cows which are not very mobile due to old age or sore feet) may only enter four to six times daily. First lactation cows usually enter more often than older cows (Voss, 1984). Occupancy time for the stalls is affected by the amount of feed being dispensed for the group and by the ratio of eating time to dawdling time. We found that dawdling time (time spent occupying the stall but not eating) was reduced by providing the cow with less protection from butting by other cows (shorter protection bars on the side of the stall) or by allowing moderate crowding of the system to occur. Twenty to 25 cows per feeder unit at 8–10 kg of concentrate each usually may be accommodated without any cows being deprived of feed due to crowding (Shrestha *et al.*, 1983).

Decision aids to assist the producer in determining the recommended concentrate allocation for each cow are starting to appear. Such aids depend on:

(1) accurate feed analysis data for the forages being fed with individual-cow ration balancing and estimated forage intake or
(2) comparison of daily milk with a standard lactation curve. Knowledge-based feeding programs for on-farm use are becoming more common and provide an additional incentive for the producer to adopt this technology (Figure 3.3).

Waste handling

Confinement housing has also resulted in a change in the way that cattle waste is handled. The use of free stalls (cubicles) has greatly reduced the amount of bedding required. Producers who changed to this system usually also had to change to handling their waste as slurry. Little bedding was mixed with the waste. This change resulted in the development of a completely new line of equipment for waste handling. Slatted floors or scraping of the cattle housing areas, often with automatic scrapers, has become widespread. Above ground storage units have become popular, especially in areas which have high water tables. Large slurry wagons are found on many farms. Injection of slurry under the soil surface to minimize odour is common in many areas.

Milking machines and associated equipment

Advances in automating and mechanizing the milking parlour account for a large part of the increased labour efficiency which has occurred in milk production. The key development was the automatic cluster detacher (remover) in the mid-1970s. Development of reliable detachers resulted in a rapid adoption of this technology. It was found that the number of milker units which could be operated satisfactorily by one person was approximately doubled by the use of automatic detachers, power entrance and exit gates and a crowd gate (Bickert, 1980). Clean-in-place washing and sanitizing of milker units (circulation cleaning) became a reality about the same time as detachers. These advancements led to the recognition that cows-per-man hour was the most appropriate unit by which milking parlour efficiency should be measured, and that parlour throughputs (cows milked/h) of 100–120 cows/h with two operators was a readily achievable goal. Even greater parlour throughputs, up to 160 cows/h can be expected with two operators and full automation (Armstrong, 1988).

The widespread acceptance of detachers, power entrance and exit gates and crowd gates in the holding area were turning points for a move away from rotary parlours to large double-sided herringbone parlours as the design of choice for new milking parlours built in the 1980s. The trigon and polygon design parlours were found to increase cow-flow by moving the cows in smaller groups relative to the standard herringbone design in the early 1980s. An advancement which led to even greater labour efficiency was the rapid exit gate. This idea was used in Israel in the early 1980s. The headgate would swing up thus allowing all of the cows to exit simultaneously by the time they had walked one cow-length. A number of manufacturers now offer rapid exit gates as an option to herringbone parlours (Armstrong, 1988).

The success of the rapid exit gates was instrumental in the acceptance of the most recent parlour design, the side-by-side stall. In these parlours milking machines are attached between the rear legs. Relative to the herringbone design, the side-by-side parlour allows the operator to walk a shorter distance and more stalls can be built in the same length building since the cows stand at right angles to the operator (O'Callaghan, 1988).

Milking parlour equipment is continuing to evolve, but some major developments may be cited. Clean-in-place washing and automatic sanitizing systems allowed operators to take full advantage of large parlours. Another piece of equipment which became a major asset for herds with mastitis being transmitted from cow-to-cow was the backflush unit. These units automatically sanitize the milker cluster unit and teat cups between each milking by a sequence of flush with water, flush with a sanitizing agent (iodine), rinse, and blow or vacuum out the residue.

Sprinklers are another automatic feature found with many milking parlours in hot climates, especially those with low humidity. The cows are 'washed' with cool water in the holding area, and dry by evaporation before entering the parlour. The effect is brief as cows rapidly heat up again after leaving the parlour, but it serves as a short respite from the heat and assists in combatting heat stress (Armstrong et al., 1988).

Automatic recording of milk yield

A major advancement which is starting to have a substantial impact on milk production technology is automatic electronic measurement and recording of milk yield. Each of the three main components for this technology – milk measurement devices, in-parlour electronic animal identification, and data recording software – required extensive development. Integrated systems with all three components developed to the stage that the entire system may truly be called automatic, started to appear in the USA around 1985. Milk meters have been available since about 1979, but most electronic animal identification systems were not able to identify more than 80–95% of the animals and the milk yield recording systems are still evolving. A main problem in the USA is how to integrate electronically measured milk yields with the national milk record system (DHI) in an efficient manner. A national feasibility trial is currently under way to test some approaches for direct electronic transfer of data from the farm to the regional milk records processing centres.

The most common method for measuring milk electronically is with a volumetric meter. However, milk is particularly difficult to measure in this way because of its tendency to foam and because the meter must operate accurately over a wide range of flow rates. Differences in milking systems between the USA and Europe have also caused some problems. Some meters apparently are more accurate under one milking environment than another. Electronic milk yields may also be obtained by the use of weighbridges to measure by weight instead of volume, instantaneous flow rates and integration over time, or by proportionate sampling.

Electronic milk meters evolved concurrently with in-parlour electronic animal identification (EID). Most major dairy equipment companies can offer every-milking recording of milk yield coupled with EID. However, many of the meters have high maintenance requirements and a relatively high initial cost.

Robot milking

Livestock engineers have dreamed and talked of a robot milker for several years, but no serious developmental effort for such a unit evolved until the early-to-mid 1980s. Two Dutch groups, the Gascoigne-Melotte Company and the combined efforts of the Dutch Institute for Agriculture Engineering (IMAG), and three commercial companies (Philips Electronics, Nedap, and Vicon) showed prototype systems at farm equipment expositions in 1988. A substantial amount of work towards a robot milker is also occurring in the UK.

While robotics will probably be incorporated into the large milking parlours in due time, the first units are designed to be integrated with automatic feed dispensers. The Vicon unit is a modular unit which will be assembled at the factory. The purchaser will be expected to provide hot and cold water, waste handling facilities, electricity and a bulk tank with a shelter for milk storage. This design will replace the milking parlour as it is known today. One unit with two side-opening stalls is expected to handle 80 cows. The robot has the capability to handle an additional two stalls constructed in a mirror image of the original unit, thus allowing one robot to milk 160 cows.

Several points are pertinent concerning the initial acceptance of the robot milker:

(1) it is almost sure to find its main market among those producers who are planning a new milking parlour. It cannot be retrospectively fitted into existing milking parlours
(2) it will greatly reduce labour but will still require substantial attention for sanitizing and checking equipment, and for making sure that all the cows are being milked
(3) it will not milk all cows automatically. Special milking arrangements will be necessary for cows whose milk is to be excluded from public sale (recently calved, mastitis, other illness), and for those cows with udders and teats outside the range of application (wide, strutting teats and low udders)
(4) milk yield/cow is likely to increase by 15–25% compared with conventional twice a day milking due to increased milking stimulation
(5) the economics of robot milking are promising if the units can be installed at a cost competitive with new milking parlours, as initial estimates indicate (Anonymous, 1988).

The equipment for the 'Automatic Milking System', as the Vicon unit is called, includes two side-opening milk stalls, a robot milker on the operator's side of the stalls, and computer controlled entrance and exit gates. The robot will move from one stall to the other on a rail to service both stalls. The milking equipment will include an electronic milk meter and a mastitis detection unit. Initial plans are for the teats to be washed in the teat cups and to discard the cleaning water and first milk.

Electronic animal identification

The key to automated data acquisition, equipment control and knowledge-based data analysis for livestock is electronic animal identification. This concept was demonstrated as early as 1970 (Broadbent, McIntosh and Spence, 1970) with the individual feeding system for animals housed in groups which is now found at cattle research stations throughout the world. The first international symposium concerning EID was held in 1976 at Wageningen, The Netherlands. This conference was followed by similar international symposia at Wageningen in 1983 and 1987 on automation in dairying.

The first widespread commercial application of EID was with automatic concentrate dispensers. Several companies developed EID systems capable of reliable operation for automatic concentrate dispensing. Most of these early units were patterned after the technologies described at the 1976 conference on EID at Wageningen (Spahr and Puckett, 1985; Kuip, 1987). The limiting feature of these EID systems for use in the milking parlour was their relatively short signal reception range, typically 10–15 cm. This range was not a problem with concentrate dispensers since each cow was required to put her head in a designated area to receive feed and the EID unit could be brought into close proximity to the reader.

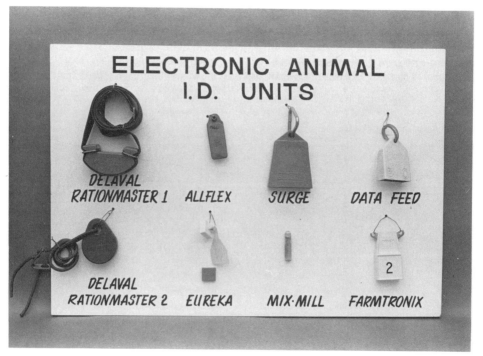

Figure 3.4 Examples of electronic animal identification units in commercial use which are being tested in the USA, 1986

Such was not the case in the milking parlour unless concentrates were fed in the parlour. Several of the EID units first designed for concentrate dispensing have been installed for automatic data acquisition in the milking parlour.

Our experience with one such system is typical (Spahr, Puckett, and Deligeersang, 1987). About 80–90% of the cows usually could be identified automatically with readers attached to the headgate section of the stall. Upon surveying several dairy farms in Europe and Israel with similar EID systems in 1986 I found a few farms that could approach 95% identification by using various techniques to improve the percentage of cows with EID units held close enough to the reader to be identified automatically. An in-parlour numeric key-pad, usually one at each stall, allowed the parlour operator to enter the identification of the cows missed by the system, so electronic acquisition of milk weights was not interrupted.

Automatic electronic identification of more than 99% of the cows in the milking parlour with a commercial system was achieved in the USA in 1985 with the release of a new portal reader and a new EID unit by Dairy Equipment Co. Since that time several additional successful systems have been introduced (Figure 3.4) (Spahr, 1988).

Several strategies to achieve more than 99% identification are found in these systems. Alfa Laval use their USA-built EID unit which was previously utilized with their 'Rationmaster II' concentrate dispenser. For in-parlour use they offer either the same reader as is used with the concentrate dispenser if identification at each stall is desired, or a portal reader with which cows are identified as they walk through an entrance.

The main advantage of a portal reader is that only one reader need be installed per group of stalls in the parlour, thus the hardware cost is minimal. A second advantage is that the design of the portal reader increases the effective range of interrogation since the electromagnetic field of the reader encloses the EID unit worn by the animal. Limitations of this strategy are that cows must stay in order between being identified and reaching the stall, and that a single missed identification could cause several cows to be identified as being at the wrong stall. These limitations have been overcome by:

(1) a high rate of correct identifications
(2) a keypad located at each stall for correction of errors
(3) an advanced software control system which minimizes the number of required manual entries.

Babson Bros. Co. (Surge brand) utilizes a unique design of a stall antenna to achieve almost 100% identification. They use a rod-shaped antenna which goes up the side of a stall. This feature effectively increases the range within which the EID units may be read and, combined with a battery powered EID unit with about 30 cm range, results in almost 100% identification. Their system is the first to eliminate a keyboard at the stall for manual entry of cow identity and to depend totally on the EID system.

The EID system offered by Nedap from Holland is sold by Germania, Westfalia, and Universal in the USA. Nedap's progress and some future developments were described by Kuip (1987) at the 1987 Dairy Automation Symposium. They have produced a small ear-tag model of their system, an approach which many people think will become widespread.

Afikim, an Israeli company, is offering an identification strategy that is different from all the other commercial companies (Carmi, 1987). Their EID unit integrates animal activity on the same sensor as identification. It is a battery powered unit which is attached to the animal's leg. It is read in the milking parlour at each milking, providing not only an identification of the animal but also data concerning her activity. The activity data are analysed via on-farm software with a desktop computer to alert the herd manager when cows are in oestrus. This system, with the Afikim milk meter, is currently under evaluation on several large dairy farms in the USA and Israel.

Udder health sensors

Maintaining udder health is one of the high priority goals of any dairy farm. Substantial progress has been made towards automatic, individual-quarter monitoring of udder health via in-line measurement of electrical conductivity of milk. Two units, one from our laboratory (Datta *et al.*, 1984; Puckett *et al.*, 1984) and one from the Dutch Institute of Agriculture Engineering (IMAG) (Rossing *et al.*, 1987), have been the subject of most of the research reports. The concept has also been studied in Germany and Japan.

Both the USA and Dutch units use in-line stainless steel electrodes to measure the conductance of the milk from each quarter before it is mixed. Electronic boards control the frequency of measurement at each stall and transmit the data to a desktop computer at discrete intervals (10 or more times per minute) where it is condensed to a single value per quarter for each milking using a real-time data collection algorithm.

The results from these units show that they are quite effective in locating infections, even at the subclinical level. However, the cost–benefit ratio of treating cows with subclinical infections has been shown to be questionable (McDermott *et al.*, 1983; Timms and Schultz, 1984), resulting in a decrease in the popularity of developing new methods for monitoring mastitis infections in the mid-1980s. The development of the robot milker gave this area of research a new justification and work is now renewed towards making automatic udder health sensors a viable commercial product.

Milk temperature sensors have also been advocated for monitoring udder health. Problems included locating the sensor where it would be unaffected by ambient temperature and finding a suitable sensor, one that was small enough and sensitive enough, but which would operate over long periods without recalibration. The most promising approach appears to be to place the sensor in the short milking tube. This approach measures just one quarter, but the measurement occurs before any air is mixed with the milk in the cluster (Maatje, Wiersma and Rossing, 1987).

Oestrus detection sensors

In contrast to a mastitis sensor, being able to predict accurately the presence of oestrus with an attached sensor would be a major accomplishment. Two approaches for electronic sensors have been shown to be feasible. A sensor which would monitor animal activity (number of steps or intensity of movement) is one approach; the second is an implanted sensor which would monitor tissue changes due to the effect of reproductive hormones.

Two companies, Afikim (Israel) and Dairy Equipment Co. (USA), have developed sensors specifically to detect oestrus by detecting the increase in cattle activity on the day of oestrus. The Afikim unit (Carmi, 1987) is a telemetric unit which is read electronically at milking time and recorded on a computer. The sensor is a part of the Afikim EID unit. It became available in commercial numbers in 1988 followed by testing on several dairies in the USA and Israel.

The unit from Dairy Equipment Co., an updated version of the unit described by Thompson and Rodrian (1983), contains its own on-board processor to analyse the activity data. Data are summarized by 2-hour intervals with an on-board memory capacity of 7.5 days. Light emitting diodes (LED) start blinking when activity for a 12-hour period exceeds activity in the same 12-hour period for the previous 3 days by at least a factor of two. A second and third LED blinks if this activity ratio exceeds 3.0 and 4.0 respectively. Thus a semiquantitative estimate of increased activity is known with a single unit.

The second concept is that of monitoring electrical conductance of reproductive tissue. The original work in this area may be credited to Aizenbud et al. (1980). The work has been extended at the US Department of Agriculture reproduction laboratory in Beltsville, MD (Lewis, Aizenbud and Lehere, 1985) and in our laboratory (Smith, Spahr and Puckett, 1986). Potential manufacturers for a commercially feasible telemetric system using this concept are known to be working on implantable sensors, but the development has proven to be more complex and slower than anticipated.

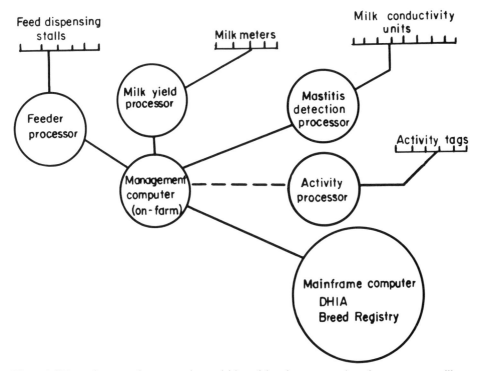

Figure 3.5 Example system for automatic acquisition of data for consumption of concentrates, milk yield, udder health and oestrous activity with telecommunications link to mainframe units in allied dairy organizations

On-farm data analysis

Cattle records are complex and challenge the best programmers. In spite of the challenges, analysis of on-farm records is an area where major advances are being made (Figure 3.5). At the end-user level, a number of database programs and decision aids are available. The mechanism for delivery of these programs varies widely, with farmer-owned cooperatives playing a strong role in some countries and multinational equipment companies being the dominant vendor in others. The cattle registry organizations, the milk recording organizations, and the agricultural extension service all play a valuable role.

On-farm record analysis has progressed through the 8-bit computer with its 64K RAM memory to today's standard of a 16- or 32-bit central processing unit (CPU), and from 4.7 MHz to 8, 12 MHz or more as the processing speed of the CPU increases. Today's desktop computer can handle artificial intelligence tasks that required specially constructed machines less than 5 years ago.

The application of this technology to on-farm analysis of records takes many forms. Database management of individual animal records is the basic starting place for most management aids designed for cattle. Cow calendar programs were widely available by the early 1980s. Such programs, with a number of advanced features, are now offered by most major milking machine manufacturers and by many agricultural software companies. The concept of automatic data acquisition

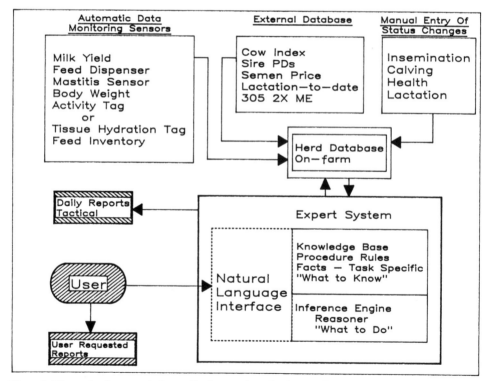

Figure 3.6 Example of automatic data collection, on-farm database and expert system analysis for dairy farms. (From Spahr, Jones and Dill, 1988)

combined with knowledge-based analysis is making substantial headway towards on-farm use, although reliable automatic sensors for mastitis detection and detection of oestrus and true expert system programs (Spahr, Jones and Dill, 1988) for cattle management (Figure 3.6) are still in the prototype stages. The spreadsheet is a specific electronic management aid which has had a significant impact on the improvement of cattle management practices. Many spreadsheet templates for specific cattle management purposes have become available with the evolution of the 16 and 32 bit CPUs and standard random access memories of 512 or 640K on desktop computers. The application of artificial intelligence to cattle production systems is emerging as a major new research area (Spahr, Jones and Dill, 1988).

Automatic acquisition of bodyweights

Another automatic data acquisition device which would benefit many cattle producers is an electronic weigh scale. A number of prototype units linked to electronic identification have been built for research farms (Adams et al., 1987; Peiper et al., 1987), and commercial versions are starting to appear. Bodyweights have substantial use in fine tuning feeding programs, and changes in bodyweight may be used as a decision aid for marketing of beef cattle or as a guideline for other management decisions. Appropriate software to acquire bodyweights automatically and to incorporate them into routine herd management programs is one of the areas where artificial intelligence approaches are expected to have an impact.

Automation in veal production

Animal welfare groups have applied substantial pressure to veal producers to change the management system for raising veal calves. The main points of the animal welfarist are that the calves should be able to move about freely, to interact with each other, and to have a comfortable place to lie.

These goals can be accomplished by housing in group pens instead of in individual pens. However, housing in group pens presents several production problems. The main problem is that calves tend to suck each other in group pens. Male calves tend to suck the prepuce of other calves resulting in drinking of urine. Such vices result in reduced growth rates, and occasionally, in death. Group housing also makes individual feeding of milk replacer more difficult than when housing is in individual pens.

Recent work in The Netherlands indicates that automation may be extended to feeding of milk replacer and may be a key to improved well-being of veal calves. Milk replacer is dispensed automatically using EID and a specially constructed dispenser unit. One type of dispenser unit (Forster) prepares a half-litre portion of milk replacer at a time. In a second type (Hokofarm) the milk replacer is mixed with water in advance and stored at 38°C in a tank. The milk is then pumped through a pipeline to the dispensing station. Results are promising (Maatje and Verhoeff, 1987) with growth rates similar to the individual feeding system. A substantial reduction in urine drinking was achieved by controlling the flow rate of milk and by the feeding of straw pellets, but no technique thus far has resulted in eliminating this vice.

Conclusions

Mechanization, automation and electronics have caused large changes in the practices employed for cattle management. Reductions in labour per unit of food produced, and in the drudgery of the required labour, are substantial. Almost all of the new practices have resulted in at least maintaining the well-being of the animal. Much of this new technology shows promise substantially to improve the management level for cattle through closer monitoring of health, reproduction, and performance responses of individual animals than has been possible previously. The advancement of electronics and their application to cattle production continues to be one of the most promising approaches to improving the efficiency of cattle production.

References

Adams, D. C., Currie, P. D., Knapp, B. W., Mauney, T. *et al.* (1987) An automated range-animal data acquisition system. *Journal of Range Management,* **40**, 256–258

Aizenbud, E., Tadmor, A., Adam, L., Atlas, D. *et al.* (1980) A preliminary report on telemetric impedance monitoring inside the vulvar tissue of ewes and heifers. *XI International Congress on Disease in Cattle.* Tel Aviv

Anonymous (1988) *The Automatic Milking System.* Vicon b.v. Neiuw Vennop, The Netherlands, 4 pp

Armstrong, D. V. (1988) Milking routine and performance of large herringbone milking parlors. In *Proceedings of the Milking Systems and Milking Management Symposium,* Northeast Regional Agricultural Engineering Service, Ithaca, NY. Publication NRAES-26, pp. 50–53

Armstrong, D. V., Wise, M. E., Torabi, M. T., Wiersma, F. *et al.* (1988) Effect of different cooling systems on milk production of late lactation Holstein cows during high ambient temperatures. *Journal of Dairy Science,* **71** (suppl. 1) 212 (abstract)

Bickert, W. G. (1980) Selecting milking parlors and mechanization. In *Proceedings of the National Milking Center Design Conference.* Northeast Regional Agricultural Engineering Service, Ithaca, NY. Publication NRAES-12

Broadbent, P. J., McIntosh, J. A. R. and Spence, A. (1970) The evaluation of a device for feeding group-housed animals individually. *Animal Production,* **12**, 245–252

Carmi, S. (1987) Automatic computerized herd management: heat detection, mastitis and health monitoring. In *Proceedings of the 3rd Symposium on Automation in Dairying.* IMAG, Wageningen, The Netherlands, pp. 18–27

Datta, A. K., Puckett, H. B., Spahr, S. L. and Rodda, E. D. (1984) Real-time acquisition and analysis of milk conductivity data. *Transactions of American Society of Agricultural Engineers,* **27**, 1204–1210

Forbes, J. M., Jackson, D. M., Johnson, C. L., Stockhill, P. and Hoyle, B. S. (1986) A method for the automatic monitoring of food intake and feeding behaviour of individual cattle kept in a group. *Research and Development in Agriculture,* **3**, 175–180

Ipema, A. H. and Rossing, W. (1987) Automatic individual feeding of dairy cattle. *Proceedings of the 3rd Symposium on Automation in Dairying.* IMAG, Wageningen, The Netherlands, pp. 41–51

Kuip, A. (1987) Animal identification. In *Proceedings of the 3rd Symposium on Automation in Dairying.* IMAG, Wageningen, The Netherlands, pp. 12–17

Lewis, G. S., Aizenbud, E. and Lehrer, A. R. (1985) Changes in electrical resistance in vulvar and uterine tissue during estrous cycles of Holstein cows. *Journal of Dairy Science,* **68** (suppl. 1), 195 (abstract)

Maatje, K. and Verhoeff, J. (1987) IVO beproeft voenomputen voor vleeskalveren in groephuisvesting. *Boerderii,* **73**, 44–45

Mattje, K., Wiersma, F. and Rossing, W. (1987) Measuring milk temperature during milking and the activity of the dairy cow for detecting sick cows and cows in oestrus. In *Proceedings of the 3rd Symposium on Automation in Dairying.* IMAG, Wageningen, The Netherlands, pp. 176–197

McDermott, M. P., Erb, H. N., Natzke, R. P., Barnes, F. D. *et al.* (1983) Cost benefit analysis of lactation therapy with somatic cell counts as indications for treatment. *Journal of Dairy Science,* **66**, 1198–1203

O'Callaghan, E. (1988) Milking systems in Europe. In *Proceedings of the Milking Systems and Milking Management Symposium.* Northeast Regional Agricultural Engineering Service, Ithaca, NY. Publication NRAES-26, pp. 11–23

Peiper, U., Maltz, E., Eden, Y., Babazada, M. *et al.* (1987) Routine automatic dairy cow's body weight measurements for the use of husbandry decisions. In *Proceedings of the 3rd Symposium on Automation in Dairying.* IMAG, Wageningen, The Netherlands, pp. 116–128

Puckett, H. B., Fernando, R. S., Spahr, S. L. and Rodda, E. D. (1984) A system for analysis of milk production and conductivity data in real-time. In *Proceedings of Agricultural Electronics – 1983 and Beyond.* American Society of Agricultural Engineers, St. Joseph, MI, pp. 508–516

Puckett, H. B., Spahr, S. L. and McCoy, G. C. (1987) Sequence control of automatic batch mixing and distribution of a total mixed ration. In *Proceedings of the 3rd Symposium on Automation in Dairying.* IMAG, Wageningen, The Netherlands, pp. 28–40

Rossing, W., Benders, E., Hogewerf, P. H., Hopster, H. *et al.* (1987) Practical experiences with real-time measurements of milk conductivity for detecting mastitis in an early stage. In *Proceedings of the 3rd Symposium on Automation in Dairying.* IMAG, Wageningen, The Netherlands, pp. 138–146

Shrestha, C. M., Spahr, S. L., Puckett, H. B. and Olver, E. F. (1983) Electronic transponder feeding system performance and animal feeding behavior for four feed availability times. *Journal of Dairy Science,* **66**, 332–340

Smith, J. W., Spahr, S. L. and Puckett, H. B. (1986) Changes in relative electrical resistance and conductivity of vaginal and vulvar tissue in relation to estrus in dairy cows. *Journal of Dairy Science,* **69** (suppl. 1), 92 (abstract)

Smith, T. R. and Prictard, D. E. (1983) An overview of recent developments in individual concentrate feeding equipment and management. In *Proceedings of the Second National Dairy Housing Conference.* American Society of Agricultural Engineers, St Joseph, MI, pp. 150–159

Spahr, S. L. (1988) Recent advances in electronic identification and data acquisition. In *Proceedings of the Milking Systems and Milking Management Symposium.* Northeast Regional Agricultural Engineering Service, Ithaca, NY. Publication NRAES-26, pp. 62–71

Spahr, S. L., Jones, L. R. and Dill, D. E. (1988) Expert systems – their use in dairy herd management. *Journal of Dairy Science,* **71**, 879–885

Spahr, S. L. and Puckett, H. B. (1985) Recent progress in the development of animal electronic identification systems. In *Proceedings of the 89th Annual Meeting of US Animal Health Association,* pp. 278–282 US Animal Health Association, Richmond, VA

Spahr, S. L., Puckett, H. B. and Deligeersang, (1987) Performance of electronic animal identification systems in the milking parlor. In *Proceedings of the 3rd Symposium on Automation in Dairying.* IMAG, Wageningen, The Netherlands, pp. 285–394

Timms, L. L. and Schultz, L. H. (1984) Mastitis therapy for cows with elevated somatic cell counts or clinical mastitis. *Journal of Dairy Science,* **67**, 367–371

Thompson, P. D. and Rodrian, J. A. (1983) Transducers for capture of activity data. In *Proceedings of the Symposium on Automation in Dairying.* IMAG, Wageningen, The Netherlands, pp. 115–126

Voss, T. A. (1984) Behavior of dairy cattle using electronic feed stalls. *M.Sc. Thesis,* University of Illinois, Urbana, USA, 62 pp

Chapter 4

New techniques in hormonal manipulation of cattle production

J. F. Roche

Successful cattle production involves the efficient conversion of minimum cost, balanced diets by healthy animals into high quality beef or milk. This requires the animals to be of the correct genetic merit for the particular production trait selected. Hence, it is vital that dairy replacements or calves destined for beef production are of high genetic merit when entering the production cycle. Thus, increased use of progeny tested bulls on an improving female population is required, which, in reality, means continued expansion of artificial insemination (AI). In addition, in the case of beef production, the number of calves produced/100 cows per annum needs to be maximized, particularly in those countries where significant numbers of calves for beef production originate from the dairy herd, due to the decrease in dairy cow numbers within the EEC. Hence, the major part of this chapter deals with current and future methods to increase reproductive efficiency of cows, with a view to producing maximum numbers of calves of the correct genetic merit, and if it were possible, of the desired sex.

However, production efficiency can also be improved by the use of various chemicals that are known to modulate different physiological processes within the body, with the net effect of increasing efficiency of production. Therefore, the hormonal manipulation of growth and lactation will also be briefly reviewed. A successful cattle production industry must produce products that are in high demand and at acceptable prices to an increasingly discriminating affluent consumer, who is faced with a myriad of choice of home-produced and imported animal products. There is also increasing consumer awareness regarding health issues particularly relating to fat content and presence of potentially harmful chemical residues in food. The use of production enhancers is under increasing scrutiny by the consumer, culminating in the current EEC ban on the use of androgens, oestrogens and progestagens as anabolic agents in meat production. Thus, the future use of chemical production enhancers within the EEC is hard to predict because decisions made regarding their use sometimes ignore the scientific data on their safety.

Reproductive efficiency in cows

Targets

In order to maximize milk production, it is necessary to have each cow in the herd produce a live healthy calf, of the correct genetic merit every year. Since the duration of pregnancy varies from 282–287 days, this results in a post-partum period of 78–83 days for the cow to become pregnant to achieve this target. Following calving, a variable period of 30–45 days is required for uterine involution to take place and ovarian cyclicity to be resumed. Conception rate also is reduced in cows that are served before days 42–50. Therefore, to attain a 365-day calving interval the cow must be detected in oestrus, served and conceive between days 40 and 83 post-partum. This means most cows only have sufficient time in this critical post-partum period to show two successive oestrous periods. Therefore, high reproductive efficiency in cows is dependent on two important factors in the post-partum period:

(1) high submission rate of cows to service from day 40 onwards
(2) high conception rate of served cows.

Submission rate to AI

The pattern of submission of cows for service will dictate the calving pattern in the herd. The target submission rate to be achieved will depend on the annual milk production pattern required within the herd, which is dictated by quotas and seasonal changes in milk price. In a predominantly spring calving herd, it is important to have the majority of oestrous cows presented for AI within a short period, once the breeding season has commenced, and a target of 75% submission rate to AI in the first 3 weeks is acceptable. In a herd with cows calving in both spring and autumn, similar targets should be set for both breeding seasons within the year, although cows that do not conceive at one time of the year can be re-bred during the subsequent breeding season with the consequence of such cows not having a 365-day calving interval.

The main factors affecting submission rate are:

(1) time of resumption of ovarian activity
(2) oestrous detection efficiency
(3) use of hormones to regulate the oestrous cycle.

(1) Resumption of ovarian activity in the post-partum period

Following calving, the uterus has to involute, both physically in relation to size and physiologically for normal maintenance of pregnancy, and the ovaries must recommence cyclical activity, ovulation and normal luteal function. Physical involution involves sloughing off of caruncular material, decrease in myometrial size, and the expulsion of exudate from the uterine lumen. During the early post-partum period, the concentration of prostaglandin (PG) $F_{2\alpha}$ is high but the role that this hormone plays in uterine involution is unclear (Edqvist, Kindahl and Stabenfeldt, 1978). The incidence of dystocia, retained placentae, and uterine infection all affect the rate of uterine involution (Marion, Norwood and Gier, 1968).

Currently, there is controversy whether or not exogenous $PGF_{2\alpha}$ administration will affect the rate of uterine involution and reduce the interval from calving to conception. Some authors have reported that a single injection of $PGF_{2\alpha}$ will improve post-partum reproductive efficiency (Young and Anderson, 1986), while others have reported no effect (Armstrong, O'Gorman and Roche, 1989). Clearly, there is insufficient understanding of the mechanism involved in involution of the uterus, and further work is required before meaningful recommendations regarding the use of hormones can be given to decrease the time required for involution. Cows that show oestrus within 40 days of calving should not be bred at this oestrus, because the conception rate is decreased until the uterus is capable of maintaining normal pregnancy rates again. The only exception to this is late calving cows in a spring calving herd, where it is important to put them in calf as soon as possible after calving. Thus, late spring calving cows should be bred as soon as possible after calving, because date of calving is the main determinant of the length of lactation, and hence quantity of milk produced per lactation.

Resumption of ovarian activity involves a variable period of growth of small follicles, which reach 8–10 mm in size. Then one follicle continues to grow and becomes the dominant follicle, inhibiting growth of other medium follicles in both ovaries. When luteinizing hormone (LH) pulses are of sufficient frequency, the dominant follicle ovulates, and an oestrous cycle of normal duration (18–24 days), of short duration (9–11 days) or of long duration (26–30 days) ensues. Generally, early in the post-partum period, oestrus does not accompany the first post-partum ovulation. The factors known to affect the interval from calving to first ovulation and oestrus are: nutrition, suckling, parity, season, the incidence of dystocia, retained placentae and uterine infection (Morrow *et al.*, 1966). Regarding nutrition, cows should be checked for body condition at drying off, and those in poor body condition given preferential feeding until calving with due regard to potential dystocia problems. During lactation, cows should be fed a sufficient quantity of balanced high quality ration to minimize weight loss. Suckling depresses the release of LH, delays follicular development and prolongs the interval from calving to first oestrus (Carruthers and Hafs, 1980). In addition, beef cows are more prone to have silent ovulations and the external manifestations of oestrus are less intense than in dairy cows, particularly in the absence of a male. The combination of poor nutrition and suckling often results in prolonged post-partum anoestrous periods in beef suckler cows, particularly those calving in autumn.

It is important to have cows not already seen in oestrus that are 6 weeks calved checked by a veterinarian to determine if cows:

(a) have resumed ovulation
(b) are still anoestrus
(c) have a specific clinical problem.

This 6-week check is an important way to improve submission rate and have most cows inseminated by days 50–55 after calving. This is an important target to achieve and is the first step in achieving a calving to conception interval of 85 and a calving interval of 365 days.

Cows that are anoestrus should be checked to determine the nature of the problem. If it is a management problem, this should be corrected. If hormonal therapy is to be considered, the two main choices are:

(a) insert an intra-vaginal progesterone releasing device (PRID) for 9–12 days and either breed cows 56 h post-removal or at detected oestrus 2–3 days after removal (Roche and Ireland, 1984). For dairy cows in reasonable body condition, over 80% will respond. However, beef suckler cows, should be given an injection of pregnant mares' serum gonadotrophin (PMSG), which has mainly follicle stimulating hormone (FSH) properties, to stimulate follicular growth and ovulation

(b) a single injection of gonadotrophin releasing hormone (GnRH) or one of its potent analogues can be given which will cause an LH/FSH surge and ovulation, if there is a large follicle present capable of ovulation (Peters and Lamming, 1984). A variable percentage will ovulate, depending on follicular status, and many cows that do ovulate will have a short ovarian cycle and show oestrus 9–12 days post injection.

(2) Oestrous detection efficiency

As already discussed, most cows have the chance of showing only two periods of oestrous behaviour between days 40–82. In addition, the mean length of oestrus is 8 hours and up to 30% of dairy cows can have a standing heat of less than 6 hours' duration (O'Farrell, 1984). The duration is shorter and the intensity of expression of oestrus weaker in suckler cows. Combined with the above problems, is the fact that conception rate to a single insemination varies from 50 to 60% (Roche et al., 1978). Therefore, the efficient detection of oestrus, where AI is being used, is the major factor affecting submission rate to AI.

Cows generally exhibit oestrous behaviour best in their normal environment, and concrete surfaces or slats are not conducive to optimum expression of oestrus. Cows have more intense interactive behaviour when there are two or more in oestrus or in the follicular (17–21) or early luteal phase (days 1–3) at the same time. Cows in the mid-luteal phase or pregnant cows have less interest in mounting oestrous cows. The most accurate behavioural sign of oestrus is standing to be mounted by another cow or a male, although other signs such as restlessness, decreased milk yield, presence of clear mucus from the vulva, red hyperaemic vulva, hair loss and abrasion of the skin of the tail head, all help in identification of the oestrous cow. Frequency of detection, with minimum time between the last check in the evening and the first check the following morning, is the single most important factor in efficient oestrous detection (O'Farrell, 1984). An ideal routine is to check every 4 h beginning at 06:00 h and ending at 22.00 h. The efficiency of oestrous detection is low on many farms (Fagan and Roche, 1986) varying from 35 to 70%, and the apparent reason for this is twice-a-day heat detection at morning and evening milking. Aids to detection of oestrus are generally not very reliable, but painting the tail head of the cow with a strip of emulsion paint 20 cm (8 inches) long and 5 cm (2 inches) wide can help (Macmillan, 1988). In general, there is no substitute for the frequent keen eye of a good stockman, in the absence of the bull.

(3) Control of oestrus

The hormonal control of oestrus in cows has recently been reviewed (Roche and Ireland, 1984) and thus the main uses will be summarized here. The two methods currently used to control oestrus and ovulation resulting in normal fertility are:

(a) use of progesterone to hold animals at the onset of the follicular phase until normal regression of the corpus luteum (CL) has occurred in all cows
(b) cause premature regression of the CL after days 5–7 by an injection of the luteolytic agent, $PGF_{2\alpha}$.

The main methods of using progesterone to control oestrus and ovulation are presented in Table 4.1. The important points to note when using progesterone are:

Table 4.1 The common methods using progesterone or progestagen to control oestrus in cattle

Method	Luteolytic agent	Duration of treatment	Onset of oestrus*	Fertility
PRID	10 mg capsule oestradiol	10–12 days	2–3 days	Normal
Ear implant of norgestomet (N)	3 mg N + 5 mg oestradiol valerate at start of treatment	9–10 days	2–3 days	Normal

* Post progesterone/progestagen removal; PRID-progesterone releasing intravaginal device

(a) to limit duration of treatment to 9–12 days
(b) use of a luteolytic agent with progesterone which can be a capsule of oestradiol adhered to the PRID, an injection of 5 mg oestradiol benzoate or valerate at the start of treatment, or an injection of $PGF_{2\alpha}$ at or within 1–2 days at the end of treatment
(c) the onset of oestrus is precise with most cows in oestrus 2–3 days after progesterone withdrawal
(d) fixed time insemination can be used in dairy cows with acceptable results
(e) about 10% of cows do not show oestrus until 4–6 days after the end of treatment and these should be re-inseminated at a detected oestrus to achieve best results
(f) this treatment will stimulate many anoestrous dairy cows to ovulate.

The main methods of using $PGF_{2\alpha}$ to control ovulation are shown in Table 4.2, and the important points to note when using it in cattle are:

(a) it is not effective from days 1 to 5 or 6 of the cycle
(b) the onset of oestrus is more variable, particularly when given at mid-cycle, which is related to the state of follicular development at the time of $PGF_{2\alpha}$ injection

Table 4.2 Methods of use of prostaglandins to control ovulation in cyclic cattle

	Treatment	When to AI
(i)	Two injections of PG 11 days apart	72 + 96 h for heifers
(ii)	Inject all cows and breed Re-inject non-responders in 11 days	At detected oestrus
(iii)	Breed cows for 7 days normally Inject remainder with PG	At detected oestrus
(iv)	Rectal examination and inject cows with CL	At detected oestrus
(v)	Inject cows with progesterone levels >1 ng/ml serum or >5 ng/ml milk	At detected oestrus
(vi)	Ultrasound and inject cows with CL	At detected oestrus

PG: prostaglandin; CL: corpus luteum.

(c) fixed time insemination should not be used in dairy cows given $PGF_{2\alpha}$ because of the spread in onset of oestrus
(d) it has no effect if given to anoestrous cows
(e) it will terminate pregnancy up to the fifth month
(f) it has limited use in suckler cows because a high percentage of them are still anoestrus 50 days after calving.

Conception rate to AI

The attainment of high conception in cows requires a healthy female to be inseminated, either naturally or artificially, with sufficient high quality spermatozoa and to allow a single capacitated, hyperactive spermatozoon to fertilize the ovum in the oviduct within 2–4 h of ovulation. The main factors affecting conception rate in cows are:

(1) the male
(2) the female
(3) management decisions.

(1) The male effect

This can initially be divided into natural or artificial service. Where natural mating is being used, the bull should be run with 30–40 cows. He should be fit and not fat, be properly fed, healthy and experienced, have normal testicular development and be capable of, and in an environment conducive to, normal sexual behaviour, mounting and copulation. Where AI is being used, the semen should have originated from a fertile bull, and have been properly collected, frozen, stored, thawed out prior to use, and the correct concentration should be used (Macmillan, 1988). The semen should be placed in the common body of the uterus with minimum stress by an experienced inseminator. Both the bull and the inseminator used can affect fertility. The spermatozoa, once placed in the female tract, must set up proper sperm reserves, particular at the utero-tubual junction and in the oviducts. They have to become capacitated and hyperactive, which takes 2–6 h in the female tract. The viable life span of bull spermatozoa varies from 12–24 h, depending on the bull used. It is currently very difficult to improve fertility of the male by hormonal therapy.

(2) The female effect

The critical factor in achieving a high conception rate is the time of AI relative to the time of ovulation, with an optimum time to breed being 12–15 h prior to ovulation. This allows sperm reservoirs to be set up, capacitation to take place and an adequate population of highly motile sperm to be present in the oviduct when the cow ovulates (Hawk, 1983). Since the cow ovulates 10–14 h following the end of oestrus, and the fertile life span of the egg is 6–12 h and that of the spermatozoa 18–24 h, the time of AI is particularly critical in cattle in order to achieve high conception rates. Increasing the conception rate by use of hormones in the female has been attempted and the approaches centre on either controlling the time of ovulation post oestrus or reducing embryo mortality by progesterone therapy. To control the time of ovulation, either GnRH or human chorionic gonadotrophin (hCG) which has LH activity, can be injected at the time of AI or service and ovulation will take place 24–28 h later (Stevenson, Schmidt and Call, 1984). This approach does not improve fertility in normal cows, but it may have an effect on cows prone to have delayed ovulation (high yielding cows ovulating within 60 days of calving or cows fed on diets deficient in β-carotene) or cows being inseminated more than three times in attempts to become pregnant. Administration of exogenous progesterone following AI has given equivocal results to-date and is not currently recommended in cows (Diskin and Sreenan, 1986).

(3) Management factors

The main management factors affecting conception rate are:

(a) interval from calving to service, with cows bred less than 50 days after calving having lower conception rates
(b) plane of nutrition, with cows in poor body condition and losing weight in early lactation having lower conception rates
(c) trace mineral problems such as excess molybdenum which induces copper deficiency causing lower conception rates
(d) presenting only oestrous cows for AI and presenting those in oestrus at the right time in relation to expected time of ovulation.

Embryo production

The ovaries of the cow contain thousands of primordial follicles, a number which is fixed at birth. Despite the fact that there are thousands of follicles present in the ovary, only one follicle usually ovulates per oestrous cycle. Once puberty has been reached, the cow will ovulate 20–40 follicles during her lifetime and from these she will produce four to seven calves. Thus, the ovary contains a large pool of unused genetic material. The two methods currently used in attempts to harvest a greater number of embryos per cow are:

(1) hormonal induction of superovulation to obtain three to ten embryos per oestrous cycle
(2) in vitro collection, maturation and fertilization of ova obtained from ovaries removed from slaughtered cattle.

(1) Superovulation of cattle

This has generally been carried out using a single injection of PMSG given (i) on day 10 and followed 2 days later with an injection of PG to cause luteolysis, (ii) given at or near the end of a short-term progestagen treatment, or (iii) given during the follicular phase of the cycle (Moor, Kruip and Green, 1984). A dose of 2000–3000 iu of PMSG can be used and there is variation in the superovulatory response obtained (0–40 ovulations) and in the number of viable embryos produced (0–8).

Factors that affect the response to PMSG are season, dose, breed, age and long half-life of PMSG (Gordon, 1982; Monniaux, Chupin and Saumande, 1983). The LH content of the PMSG can induce premature resumption of meiosis (Moor, Osborn and Crosby, 1985) so that aged ova are ovulated, and if these are fertilized they produce degenerating embryos which do not survive. The long half-life of

PMSG may stimulate follicular development after ovulation and results in rescue of some atretic follicles which fail to ovulate, resulting in very high concentrations of oestradiol on days 2–6 of the next cycle, which appears to be unfavourable to embryo survival. The use of PMSG antibodies to neutralize the excess PMSG given near oestrus will decrease the number of unovulated follicles but equivocal results have been obtained regarding an increase in number of viable embryos obtained.

The current alternative to PMSG is the use of purified FSH, which must be repeatedly injected because it is rapidly cleared from the blood. A dose of 32 mg FSH in cows and 24 mg in heifers is given during the mid-luteal phase as two injections per day for 4 days in decreasing doses with PG given on the third day of the treatment regimen (Roche and Ireland, 1984). The FSH preparation used should have a minimal content of LH. Attempts to develop slow release systems for administration of FSH have not yet been successful. The ovulatory response from such an FSH regimen varies from 12 to 18 ovulations with two to seven embryos produced per donor treated.

(2) In vitro fertilization (IVF)

Neither ejaculated bull spermatozoa nor ova recovered from follicles on ovaries obtained from cows at unknown stages of the reproductive cycle are capable of fertilization without first undergoing maturation. Spermatozoa can be capacitated in vitro by addition of heparin, having first isolated highly motile sperm by a swim up technique (First and Parrish, 1988). Ova can be matured either completely in vitro in specialized media or initially in vitro and then in vivo by transfer to the oviduct of a rabbit or ewe. Ova are recovered from follicles on ovaries obtained at cattle abattoirs, transferred to the laboratory within 1–2 h, vesicular follicles dissected and oocytes liberated by puncture (Gordon et al., 1987).

The oocytes are cultured in tissue 199 medium supplemented with 20% serum from an oestrous cow and extra cumulus cells for 24 h at 39°C using a flux culture system and a gas phase of 5% CO_2 in air. Oocytes and sperm are mixed in microdroplets of modified Tyrodes medium under oil, and fertilization takes place within 12–18 h. In one study (Gordon et al., 1987) 1280 primary oocytes were matured in vitro, 1071 were exposed to capacitated sperm, 975 were placed in sheep oviducts, 638 were recovered, 414 showed evidence of fertilization and 248 embryos were produced of which 153 were of freezable quality.

Despite the major advances that have been made in cattle IVF, much work remains to be undertaken to achieve acceptable pregnancy rates following transfer. However, these developments will come and already commercial interest has developed in IVF, so that within a few years, farmers may be able to get cows pregnant with embryos of high genetic merit produced by IVF.

Embryo transfer

Cattle embryos when transferred at day 6 or 7 give best results and recipients need to be synchronized within a day as regards stage of cycle for transfer purposes. Embryos are placed in 0.5 ml plastic straws and inserted into the uterus by means of a cattle AI gun as close as possible to the utero-tubal junction. The methods of transfer commonly used on day 7 are:

(1) single egg transfer in the ipsilateral uterine horn to establish pregnancy

(2) double egg transfer, one to each uterine horn in attempts to establish twin pregnancy
(3) single egg transfer to the contralateral horn of the bred recipient in attempts to establish twin pregnancy.

The pregnancy rate is 10–30% higher when the embryo is transferred to the horn ipsilateral to the CL, but the reasons for this are not clear, although it is known that there is a local effect of each uterine horn on the ipsilateral ovary.

Thus, embryo transfer can be used either to get cows pregnant with specific genetically selected embryos or to increase the twinning rate by transfer of embryos to the contralateral uterine horn in either the bred (single embryo transfer) or unbred (twin embryo transfer) recipient at the same stage of the cycle. Factors that affect the success of embryo transfer need to be clearly elucidated but the potential exists that embryo transfer can supersede AI as the method of getting cows pregnant. However, the major advance required for widespread use of embryo transfer is the prediction of the sex of the embryo. This would have a major effect in the efficiency of production of meat and milk by obtaining males and females for breeding from elite females, and males for meat production from the remainder of the breeding females.

Developing reproductive technologies

The developing techniques to improve reproductive efficiency in the future can be summarized as follows:

(1) sexing of embryos and/or sperm without loss of fertility
(2) mass production of embryos of desired genetic potential by IVF
(3) mass production of transgenic animals with the genetic ability to increase efficiency of the desired production trait of economic benefit
(4) cloning of high quality embryos to allow repeatable controlled production of a specific gene or genes
(5) predictable methods (a) to detect oestrus without the need to observe cows on a daily basis, (b) to check for pregnancy and (c) to induce single or multiple ovulations in cattle for twinning or embryo production purposes.

Hormonal regulation of growth

The precise endocrine, metabolic and genetic mechanisms controlling growth, feed efficiency and carcass composition of cattle are not clear. The expression of maximum growth rate is dependent on genetic potential and the provision of optimal nutritional and environmental conditions to disease free animals. Despite our lack of clear understanding of the molecular mechanisms involved in regulating growth, it is possible to manipulate growth of cattle by various means:

(1) use of chemical messengers which bind to specific receptors in the target tissue, thereby initiating a series of biochemical events within the cell, resulting in increased protein synthesis
(2) use of antimicrobial compounds which change the population of microorganisms in the alimentary tract of healthy animals, resulting in the improvement of animal performance.

The chemical messengers that affect growth can be divided into various classes based on chemical structure.

Steroids or compounds with steroidal activity

These include the sex steroid hormones such as oestradiol, progesterone and testosterone; the synthetic steroid hormones such as trenbolone acetate (Finaplix, Roussel-Uclaf Ltd., France); and the non-steroidal chemicals which mimic the effects of steroids such as diethyl stilboestrol (DES) or zeranol (Ralgro, IMC Ltd., US). Since these compounds have been banned by the EEC for use in cattle to stimulate growth, they will not be discussed further, but their use has been recently reviewed (Roche and Quirke, 1986).

Protein or peptides

Included in this class of chemicals are growth hormone (GH), growth hormone releasing factor (GRF) and thyroid hormone releasing factor (TRF). The major current potential hormone to affect growth rate in this class is growth hormone, which is a single chain peptide of 191 amino acids, with two intrachain disulphide bridges (Hart and Johnsson, 1986). Its structure is species specific, with greater homology of structure evident in non-primates. It has a short half-life (20–30 min), is not orally active and is rapidly metabolized and cleared by the gut, liver and kidney. The secretion of GH from the pituitary gland is regulated by the secretion of GRF and somatostatin, which is a potent inhibitor of GH release. It is a peptide containing 14 amino acids, and it is produced from the hypothalamus and various other organs in the body, including the intestine. Growth hormone releasing factor is a polypeptide containing 44 amino acids and it is rapidly metabolized by the liver and kidney. Various synthetic analogues of GRF have been prepared, which are more potent in stimulating GH release but none is currently commercially available.

Growth hormone is released in an episodic manner from the pituitary. It binds to hepatocyte membrane-bound receptors whose numbers are affected by insulin. This binding stimulates the production of a GH intermediate, called somatomedin, also known as insulin-like growth factor (IGF-1), which mediates many of the effects of GH. Insulin-like growth factor-1 stimulates protein synthesis, amino acid transport and uptake, and binds to receptors in muscle and bone; it also has mitogenic properties and its production is affected by nutritional status (Peel and Bauman, 1987). Growth hormone stimulates (i) sulphate incorporation into cartilaginous tissue, (ii) amino acid transport and uptake into bone and muscle, (iii) lipolysis, (iv) glucose uptake into tissue. Thus it has a protein-sparing effect.

When administered to farm animals GH will increase growth rate (5–8%), feed conversion efficiency and lean meat in the carcass. However, it is not a strong anabolic agent in cattle, nor does it seem to have dramatic effects on carcass composition. Hence, it is unlikely to be a major contender for widespread practical use as an anabolic agent in beef cattle in the near future.

β-Adrenoceptor agonists

These are analogues of adrenalin or noradrenalin, and they bind to specific β-2 receptors present on adipocytes, blood vessels, gastrointestinal tract and muscle

Table 4.3 Effects of including cimaterol in the diet of steers*

	Control	Cimaterol	Ratio control:cimaterol
Growth rate (g/day)	816	1064	1.30
Food efficiency (kg gain/kg food)	0.085	0.111	1.31
Carcass weight (kg)	340	379	1.11
Kill-out (%)	55.6	60.2	1.08
% separable fat	22.1	13.8	0.62
%lean	59.8	69.7	1.17
L. dorsi area (cm^2)	67.2	94.6	1.41
Kidney knob plus channel fat (kg)	4.81	2.68	0.56

* Cimaterol fed (49 mg/head per day) in the diet of Friesian steers during a 13-week finishing period.

From Hanrahan et al., 1986.

cells. The characteristics- of a specific β-agonist depend on binding affinity, specificity of binding in target tissue, and biological activity of the specific analogue (Timmerman, 1987). They increase muscle accretion, decrease fat content by lipolytic effects and perhaps by decreasing lipogenesis, and they also have marked effects on the cardiovascular system. In steers, lean meat content is increased by 12%, fat content decreased by 7% and killing-out percent increased by 2–3% (Table 4.3). There are few published data on residues, toxicity and safety of these compounds in the literature and none is currently cleared for use in beef production, despite their dramatic effects in increasing meat production.

Hormonal stimulation of lactation

Although the hormonal regulation of lactation is not clearly understood, it is known that GH will dramatically increase milk yield by 20–40%, with feed intake adjusting upwards over the first 6–10 weeks to compensate for this extra production, but overall feed efficiency is improved, and chemical composition of milk is not changed (Peel and Bauman, 1987). A daily dose of 27 mg GH, of biological or biotechnological origin is effective, and slow release formulations are being tested with a potential effective life-span of 2–4 weeks. In the initial studies carried out, bodyweight, incidence of metabolic diseases, general health and welfare of cows were not adversely affected (Whitaker et al., 1988). The protein and fat content, lactose concentration, vitamin and mineral levels and somatic cell counts of milk obtained from control and treated cows were not different. Some long-term studies have been carried out and no adverse effects on health or body condition have been reported (Peel and Bauman, 1987), although there is a suggestion that fertility may be adversely affected (Phipps et al., 1988). However, increased milk production per se is known to decrease conception rate, but there are no consistent reports in the literature indicating any effects on reproduction other than that related to higher production. However, since it is known that GH binds to granulosa cells to increase production of IGF-1, which may affect FSH action within the follicle, further detailed studies on the effects of GH, both on general reproductive parameters and on detailed endocrine events within the ovary are required before firm well-founded conclusions can be drawn. Regarding toxicological effects, GH being a protein is not orally active and its biological effects are species specific, but further testing is required in the target species regarding toxicology, drug tolerance and clinical safety before this drug will be licensed for commercial use.

Therefore, based on currently available data, it can be concluded that GH is an effective method to increase milk yield, without having undue adverse effects on milk composition, general health and welfare of cows. If it is licensed for use, the question of the most effective method to use it will be raised. Currently, it appears that it should initially be used in cows confirmed pregnant, for management of milk yield to produce milk most efficiently to meet EEC quotas.

Conclusions

Hormones can be used effectively to regulate reproduction, growth and lactation efficiency in cattle. In reproduction, their main use is to induce oestrus and ovulation in anoestrous cows or to control the time of ovulation in cyclic cattle. They can also be used to induce superovulatory responses in cattle and allow viable embryos to be collected non-surgically at 7 days after breeding. Embryo transfer has great potential in the future but more understanding is required of factors that affect single or twin pregnancy rates after transfer. The problem of a cheap and simple supply of cattle embryos of correct breed and sex is still under active research, and progress is being made in IVF technology. However, this technology has still to undergo a great deal of refinement before it will be commercially viable.

Growth and lactation output and efficiency can be increased dramatically by appropriate chemical messengers, but the real question is the acceptability of food produced by such technology by the consumer. Despite the scientific evidence proving safety in the case of steroid hormones, political widsom led to their banning as anabolic agents in meat production in the EEC. The fate of GH and β-agonists as production enhancers remains to be determined and their future commercial value is hard to predict at this time. The real lesson to be learned is the need for effective consumer and political education on safety of use and acceptability of meat and milk produced by this new era of technological development. It will be an uncompetitive, inefficient European cattle industry that stands still in the face of developing technology.

References

Armstrong, F., O'Gorman, J. and Roche, J. F. (1989) The effects of prostaglandin administration on reproductive performance in dairy cows. *The Veterinary Record*, (in press)

Carruthers, T. D. and Hafs, H. D. (1980) Suckling and four-times daily milking: influence on ovulation, estrus and serum luteinizing hormone, glucocorticoids and prolactin in postpartum Holsteins. *Journal of Animal Science*, **50**, 919–925

Diskin, M. G. and Sreenan, J. M. (1986) Progesterone and embryo survival in the cow. In *Embryonic Mortality in Farm Animals*, (ed. J. M. Sreenan and M. G. Diskin). Martinus Nijhoff Publishers, for The Commission of the European Communities, pp. 142–158

Edqvist, L-E., Kindahl, H. and Stabenfeldt, G. (1978) Release of prostaglandin $F_{2\alpha}$ during the bovine peripartal period. *Prostaglandins*, **16**, 111–119

Fagan, J. G. and Roche, J. F. (1986) Reproductive activity in postpartum dairy cows based on progesterone concentrations in milk or rectal examination. *Irish Veterinary Journal*, **40**, 124–131

First, N. L. and Parrish, J. J. (1988) Sperm maturation and *in vitro* fertilization. In *Proceedings of the 11th International Congress on Animal Reproduction and Artificial Insemination, Dublin.* **5**, 161–168

Gordon, I. (1982) Synchronization of estrus and superovulation in cattle. In *Mammalian Egg Transfer*, (ed. C. E. Adams). CRC Press, Inc., Boca Raton, Florida, pp. 63–80

Gordon, I., Lu, K. H., Gallagher, M. and McGovern, H. (1987) Production of cattle embryos by *in vitro* and *in vivo* culture. *Animal Production*, **46**, 498 (abstract)

Hanrahan, J. P., Quirke, J. F., Bomann, W., Allen, P. *et al.* (1986) β-agonists and their effects on growth and carcass quality. In *Recent Advances in Animal Nutrition*, (ed. W. Haresign). Butterworths, London, pp. 125–138

Hart, I. C. and Johnsson, I. D. (1986) Growth hormone and growth in meat-producing animals. In *Control and Manipulation of Animal Growth*, (ed. P. J. Butterly, D. B. Lindsay and N. B. Haynes). Butterworths, London, pp. 135–159

Hawk, H. W. (1983) Sperm survival and transport in the female reproductive tract. *Journal of Dairy Science*, **66**, 2645–2660

Macmillan, K. L. (1988) Maximizing the use of AI in cattle. In *Proceedings of the 11th International Congress on Animal Reproduction and Artificial Insemination*, Dublin. **5**, 265–275

Marion, G. B., Norwood, J. S. and Gier, H. T. (1968) Uterus of the cow after parturition: factors affecting regression. *American Journal of Veterinary Research*, **29**, 71–75

Monniaux, D., Chupin, D. and Saumande, J. (1983) Superovulatory responses of cattle. *Theriogenology*, **19**, 55–81

Moor, R. M., Kruip, Th. A. M. and Green, D. (1984) Intraovarian control of folliculogenesis: limits to superovulation? *Theriogenology*, **21**, 103–116

Moor, R. M., Osborn, J. C. and Crosby, I. M. (1985) Gonadotrophin-induced abnormalities in sheep oocytes after superovulation. *Journal of Reproductive Fertility*, **74**, 157–172

Morrow, D. A., Roberts, S. J., McEntee, K. and Gray, H. G. (1966) Postpartum ovarian activity and uterine involution in dairy cattle. *Journal of the American Veterinary Medical Association*, **149**, 1596–1609

O'Farrell, K. J. (1984) Oestrous behaviour, problems of detection and relevance of cycle lengths. In *Dairy Cow Fertility*, (ed. R. G. Eddy and M. J. Ducker). British Veterinary Association Editing Services, London, pp. 47–59

Peel, C. J. and Bauman, D. E. (1987) Somatotropin and lactation. *Journal of Dairy Science*, **70**, 474–486

Peters, A. R. and Lamming, G. E. (1984) Reproductive activity of the cow in the post-partum period. II. Endocrine patterns and induction of ovulation. *British Veterinary Journal*, **140**, 269–280

Phipps, R. H., Weller, R. F., Austin, A. R., Craven, N. *et al.* (1988) A preliminary report on a prolonged release formulation of bovine somatotrophin with particular reference to animal health. *The Veterinary Record*, **122**, 512–513

Roche, J. F. and Ireland, J. J. (1984) Manipulation of ovulation in cattle. In *Proceedings of the 10th International Congress on Animal Reproduction and AI*, Illinois. **4**, 9–17

Roche, J. F. and Quirke, J. F. (1986) The effects of steroid hormones and xenobiotics on growth of farm animals. In *Control and Manipulation of Animal Growth*, (ed. P. J. Buttery, D. B. Lindsay and N. B. Haynes). Butterworths, London, pp. 39–51

Roche, J. F., Sherington, J., Mitchell, J. P. and Cunningham, J. F. (1978) Factors affecting calving rate to AI in cows. *Irish Journal of Agricultural Research*, **17**, 147–157

Stevenson, J. S., Schmidt, M. K. and Call, E. P. (1984) Gonadotrophin-releasing hormone and conception of Holsteins. *Journal of Dairy Science*, **67**, 140–145

Timmerman, H. (1987) β-Adrenergics: physiology, pharmacology, applications, structures and structure-activity relationships. In *Beta-Agonists and their Effects on Animal Growth and Carcass Quality*, (ed. J. P. Hanrahan). Elsevier Applied Science, London, pp. 13–28

Whitaker, D. A., Smith, E. J., Kelly, J. M. and Hodgson-Jones, L. S. (1988) Health, welfare and fertility implications of the use of bovine somatotrophin in dairy cattle. *The Veterinary Record*, **122**, 503–505

Young, I. M. and Anderson, D. B. (1986) Improved reproductive performance from dairy cows treated with dinoprost tromethamine soon after calving. *Theriogenology*, **26**, 199–208

Chapter 5

New techniques in cattle breeding

J. W. B. King

The intention in this chapter is to discuss new developments in the selective improvement of both beef and dairy cattle. The details of many of the reproductive techniques which are employed are discussed in Chapter 4 and attention will be focused on their application with some speculation as to how the growing repertoire of techniques might be used in the future.

It will be convenient to discuss the impact of new techniques in various categories relating to reproduction, statistical methods and developments in biotechnology (and in particular genetic manipulations).

Reproductive techniques

Artificial insemination

Although the artificial insemination (AI) of cattle has been an available technique for many years, in terms of applications it is still under-exploited. In dairy cattle breeding the widespread use of AI breaks down traditional breeding structures based on individual herds, but in many countries these still persist with beef cattle. The use of AI on beef cattle, even to a modest extent, is useful in providing genetic connections between herds and allowing for better evaluations of breeding merit – a topic to be discussed below.

The freezing of semen has been a valuable technique allowing for both nominated matings and for international trade in semen. It is worth noting, however, that the technique can be over-exploited from a national standpoint because of the loss of spermatozoa in the freezing and thawing process. With a progeny testing scheme for dairy bulls, better exploitation of the best proven sires could often be obtained by the use of fresh rather than frozen semen resulting in an economy in the number of bulls used and an increase in their average merit.

The advantages of using fresh semen in the commercial situation have been clearly recognized in New Zealand and other countries are beginning to recognize this opportunity offered by not using a technique where it is not essential.

Artificial insemination still awaits one further development and that is the sexing of semen without serious deterioration in fertilizing ability. Despite many claims made the only convincing and confirmed reports of methods which separate the X and Y sperm have involved the killing of that sperm in the process. The development of such a technique would be a prize indeed, but some of the ways in which the method might be exploited turn out to be rather unexpected. Taylor,

Theissen and Moore (1986) have shown that in beef production the use of such sexed semen might well be for the production of once-bred heifers in a self-sustaining breeding scheme rather than for the production of male calves with superior growth rates.

Aside from its value for production purposes sexed semen would, however, have further repercussions. Dual purpose breeds would lose an advantage that they have at present and specialized dairy breeds would gain. The method could be a great asset to those minority dairy breeds which do not now prosper because surplus pure bred male calves are not found suitable for beef production.

Embryo transfer

The commercial incentive of producing for sale animals of breeds which were temporarily in short supply has been largely responsible for the development of embryo transfer from a laborious laboratory method into a routine non-surgical technique that can be used in the field. While still useful for such purposes of breed substitution, the major impact in the future will almost certainly be for within breed improvement.

For dairy cattle, improvement schemes have in the past almost exclusively been modelled on field progeny testing methods using AI. Initial calculations showing what might be achieved by using embryo transfer in the breeding of young bulls were disappointing in showing only modest improvements (e.g. Cunningham, 1976). For dual purpose breeds using performance testing for beef and progeny testing for dairy traits, the gain was rather greater (Kräusslich, 1976), but still not spectacular. It was not until Nicholas (1979) hit upon a new method of using embryo transfer that the potentials of the method were appreciated. The essential feature of the new scheme was to carry out multiple ovulation and embryo transfer (MOET) on all the heifers in the dairy herd before they had commenced their first lactations. In this way once the first lactation records on a heifer were complete, replacement families were immediately available if required. The success of the alternative breeding method was dependent on the acceptance of a less accurate measure of breeding merit but was more than compensated for by a great reduction in the generation interval.

Further calculations by Nicholas and Smith (1983) showed that some compromise methods were also attractive, e.g. when heifers were allowed to complete one lactation before embryo transfer was started. This latter method, which was named the 'adult' method as against the 'juvenile' described previously, had attractions in being more economical and less dependent on the ability to carry out embryo collection from young heifers. However, some corrections to the original calculations given by Woolliams and Smith (1988) show the juvenile method to be the method of choice where feasible. Both juvenile and adult schemes use family selection methods and introduce problems of increased rates of inbreeding which have to be faced. These problems may be met in part by keeping more than one bull per sibship but are essentially inherent in such selection methods.

The structure of breeding herds was radically changed by the introduction of artificial insemination at an earlier date, and MOET schemes seem likely to introduce equally revolutionary changes at the present time. Rather than employing many different herds for progeny testing, MOET schemes make it possible to concentrate recording and selection in a few nucleus herds. The advantages of such concentration of effort had been noted earlier by Hinks (1978), but the introduction of embryo transfer makes such a system self-replenishing. In nucleus herds much tighter management can be introduced and records taken that would not be feasible in a large field scheme. Measurement of food intake makes it possible to introduce some consideration of input costs into selection decisions.

Bulls from MOET schemes would undoubtedly be progeny tested in the field by existing methods. To wait for such progeny results will, however, undermine the basic generation interval advantage of MOET schemes. Thus some hybrid schemes where nucleus herds to breed young bulls are combined with field progeny tests have been suggested (Colleau, 1985; 1986; Christensen and Liboriussen, 1986), but will not compete in the long term with MOET schemes.

Extensive work is being carried out in modelling MOET schemes to see the advantages which might reasonably be expected from them. The potential advantages are great, although they tend to be dissipated in very small populations (Juga and Maki-Tanila, 1987). Not only are there selection problems arising from the small numbers involved but also sacrifices in potential gains caused by the need to avoid too rapid rates of inbreeding.

For beef cattle improvement the advantages of MOET schemes were realized at a much earlier time (Land and Hill, 1975). Because many beef characteristics can be recorded in both sexes, individual rather than family selection can be practised with the consequence that inbreeding complications are very much reduced. Despite the big advantages in rate of genetic advance (up to 80% for growth rate) for a MOET breeding scheme the progression of these schemes into practice has been extremely slow. The reason for this is probably non-genetic resulting from the fact that embryo transfer has been an expensive process and potential developers of such schemes have realized that it would not be easy to recoup their investment. Recent developments in the *in vitro* culture and fertilization of follicular oocytes (Lu *et al.*, 1987) will certainly reduce the costs of ova transplantation and perhaps find application in a beef MOET scheme, although many breeders might find it strange to view a breeding herd with no adult cows.

Other reproductive technologies

There is an increasing understanding of early mammalian development and with it knowledge of the ways in which further manipulations might be possible. Some ideas which seemed attractive, such as the effective mating of males to males by the fusing of two male pronuclei, would form the basis of an attractive breeding programme (Van Raden and Freeman, 1988). Unfortunately, experiments with mice (Surani, Barton and Norris, 1987) show that this is not immediately possible, although the attempt in itself has thrown up some quite fundamental discoveries on the process of early development.

One technique which appears more promising is that of nuclear transplantation leading to the possibility of producing very large clones of genetically identical animals (Willasden, 1986; Marx, 1988). Some of the genetic consequences of breeding clones for commercial production have been examined by Nicholas and Smith (1983) who show that very great improvements could reasonably be expected. With clones it becomes feasible to evaluate performance characteristics in very much greater detail so providing information on food requirements, disease susceptibility, temperament etc. Widespread adoption of a clonal breeding system would change dairy cattle breeding completely. Thus we might look forward to a commercial cow population made up of several clones, almost certainly of crossbred animals. These would be bred from quite small populations of purebreds, bred presumably by old fashioned MOET systems and forming the nucleus stocks for the future.

Improving reproductive performance by breeding

For both beef and dairy cattle improvements in their natural reproductive performance would be advantageous. One of the most obvious ways of doing so would be to increase the rate of spontaneous twinning. This possibility has been the subject of data analyses in many countries with the conclusion that twinning is a heritable characteristic of the cow, but with a low heritability. To speed up improvement, investigators in America, Australia, France and New Zealand have collected together individual cows with records of repeated multiple births (*see* review by Morris and Day, 1984). These cows show higher than average twinning rates and produce daughters that perform less well than their mothers but at rates distinctly above normal. The problem is to find bulls of a merit equal to that of the cows and while this may be possible with field records on AI bulls that have large progeny groups, the problem remains for subsequent generations of the selected herd. Measurement of ovulation rate which appears to have a higher heritability may be useful, although the limiting factor in achieving high twinning rates on a herd basis will be embryo survival.

Developments in statistical methodology

At the same time as the revolution in biological techniques available to the cattle breeder, there have been developments in biometrical methodology which, in their way, are no less startling. The introduction and refinement of mixed model methods that have become known as best linear unbiased predictors (BLUP) has provided a powerful means of estimating breeding merit. In summary, the method makes an approach to real life situations in allowing for the very unequal distribution of the progeny of a sire across herds, for the continual selection of animals in herds and provides a means of estimating the genetic merit of individuals bred from overlapping generations. Although originally requiring the computing power of main frame computers for their implementation many methods are now being adapted for personal computers and therefore become that much more readily available. New methods have also been developed to deal with traits that are of an all-or-none kind such as mortality, and for traits which cannot be measured but can be allocated to various descriptive categories.

Although well suited to the analysis of extensive field data, BLUP methods will be no less valuable in coping with some of the complexities of breeding nucleus herds by MOET schemes. Thus there will inevitably be problems of animals drawn from different populations, overlapping generations and incomplete records to be coped with.

Impact of biotechnology

Developments in biotechnology will have an impact on cattle breeding at all levels. Some reproductive techniques, usually categorized as biotechnology, have already been discussed, but they are few in comparison to the many possibilities already being explored in the general field of biotechnology. On the one hand we can look forward to improved diagnostics for some diseases and to improved vaccines for protection against others. On the other hand we can debate whether methods of gene transfer will enable us to incorporate genes for disease resistance or even effectively to pre-vaccinate animals at birth.

Indirect effects of biotechnology

To the extent that biotechnological products are accepted into general practice it will be necessary for improvement methods to operate against this background. The most pressing problem of this kind is that posed by the possibility that the administration of bovine somatotrophin (BST) might become an approved husbandry technique to increase milk production. With very convincing evidence that this product is effective the incentive to use it is very great. For those using field records for selection purposes the problems created will be immense. Even given complete recording of when and for how long the product was administered the complexity of the records generated will become extremely difficult to interpret. Where financial benefits are attached to high yields the pressure for incomplete reporting will also be high.

For field progeny testing schemes various remedies suggest themselves. Owners of test daughters might be paid to use BST routinely, or alternatively not to use it under any circumstances – in either case the remedy would be an expensive one. Another remedy might be to use only the early part of the lactation record for breeding purposes since BST is unlikely to be administered during this period. This would be a cheap solution to the problem but not one likely to be of appeal because of possible long-term consequences of selecting only on part record.

Those running MOET schemes may avoid these problems but still have to face-up to the strategic question of whether or not BST administration is likely to become standard practice and therefore whether or not it could be used appropriately in their nucleus herd.

For those running beef cattle improvement schemes, similar questions are likely to arise through uncertainty about the adoption of other techniques and products such as the use of β-agonists or immunization procedures.

Direct effects of biotechnology

At the centre of biotechnology are the mainstream developments directly attributable to new techniques in the manipulation of DNA. Methods for gene transfer are available in mammals and while not without its drawbacks the direct injection of the male pronucleus appears with practice to be usable as a routine. While more sophisticated gene transfer methods are under development, the immediate problems for application are of a different kind. First, there are problems in the choice of genes for transfer, their identification where not known, and the choice of a suitable control system of promoters and enhancers to be included in the programme for transfer. Some discussion of the problem is given by Simonds and Land (1987). Secondly, problems arise that, with the present method of microinjection, inserted genes may apparently integrate at random in the genome and usually in tandem multiple copies. Each transgenic animal has to be regarded as a unique event and the family bred from it fully evaluated for performance traits. Some consideration on the use of transgenics in livestock improvement has been made by Smith, Meuwissen and Gibson (1987) showing that many options are open that could enhance the rate of genetic improvement, but that much additional testing and evaluation work will be required.

In the longer term it would be my opinion that we can look forward to transgenic dairy cows with, for example, additional growth hormone genes built in and switched on when required by a simple feed additive, and to the milk they produce having already been altered to the needs of the consumer. The big debate is perhaps whether the consumer will find such dramatic changes acceptable – but that is another story.

References

Christensen, L. G. and Liboriussen, T. (1986) Embryo transfer in the genetic improvement of dairy cattle. In *Exploiting New Technologies in Animal Breeding* (ed. C. Smith, J. W. B. King and J. C. McKay). Oxford University Press, Oxford, pp. 37–46

Colleau, J. J. (1985) Genetic improvement by embryo transfer within selection nuclei in dairy cattle. *Génétique, Sélection et Évolutión*, **17**, 499–538

Colleau, J. J. (1986) Genetic improvement by ET within an open selection nucleus in dairy cattle. *Proceedings of the 3rd World Congress on Genetics Applied to Livestock Production*, Lincoln, Nebraska, USA, pp. 127–132

Cunningham, E. P. (1976) The use of egg transfer techniques in genetic improvement. In *Proceedings of the EEC Seminar on Egg Transfer in Cattle*, (ed. L. E. A. Rowson), pp. 345–353

Hinks, C. J. M. (1978) The development of nucleus herd selection programmes in dairy cattle breeding. *Zeitschrift für Tierzüchtung und Züchtungsbiologie*, **94**, 44–54

Juga, J. and Maki-Tanila, A. (1987) Genetic change in a nucleus breeding dairy herd using embryo transfer. *Acta Agriculturae Scandinavica*, **37**, 511–519

Kräusslich, H. (1976) Application of superovulation and egg transplantation in AI breeding programmes for dual purpose cattle. In *Proceedings of the EEC Seminar on Egg Transfer in Cattle*, (ed. L. E. A. Rowson), pp. 333–347

Land, R. B. and Hill, W. G. (1975) The possible use of superovulation and embryo transfer in cattle to increase response to selection. *Animal Production*, **21**, 1–12

Lu, K. H., Gordon, I., Boland, M. P. and Crosby, T. F. (1987) *In vitro* fertilisation of bovine oocytes matured *in vitro*. British Society of Animal Production. Winter Meeting 23–25 March, 1987, Grand Hotel Scarborough Programme and Summaries. Paper no. 28, 2pp

Marx, J. L. (1988) Cloning sheep and cattle embryos. *Science*, **239**, 463–464

Morris, C. A. and Day, A. M. (1984) Potential for genetic twinning in cattle. *Proceedings of the 3rd World Congress on Genetics Applied to Livestock Production*, Lincoln, Nebraska, USA, XI, 14–29

Nicholas, F. W. (1979) The genetic implications of multiple ovulation and embryo transfer in small dairy herds. *Proceedings of the 30th Annual Meeting of the EAAP*, Harrogate CG1.11.y

Nicholas, F. W. and Smith, C. (1983) Increased rates of genetic change in dairy cattle by embryo transfer and splitting. *Animal Production*, **36**, 341–353

Simonds, J. P. and Land, R. B. (1987) Transgenic livestock. *Journal of Reproduction and Fertility*, Suppl. **34**, 237–250

Smith, C., Meuwissen, T. H. E. and Gibson, J. P. (1987) On the use of transgenes in livestock improvement. *Animal Breeding Abstracts*, **55**, 1–10

Surani, M. H., Barton, S. C. and Norris, M. L. (1987) Experimental reconstruction of mouse eggs and embryos: an analysis of mammalian development. *Biology of Reproduction*, **36**, 1–16

Taylor, St. C. S., Theissen, R. B. and Moore, A. J. (1986) Single sex beef cattle systems. In *Exploiting New Technologies in Animal Breeding*, (ed. C. Smith, J. W. B. King and J. C. McKay). Oxford University Press, Oxford, pp. 37–46

Van Raden, P. M. and Freeman A. E. (1988) Potential genetic gain from producing bulls with only sires as parents. *Journal of Dairy Science*, **68**, 1425–1431

Willasden, S. M. (1986) Nuclear transplantation in sheep embryos. *Nature (London)*, **320**, 63

Woolliams, J. A. and Smith, C. (1988) The value of indicator traits in the genetic improvement of dairy cattle. *Animal Production*, **46**, 333–346

Chapter 6

New techniques in feed processing for cattle

A. S. El-Shobokshy, D. I. H. Jones, I. F. M. Marai, J. B. Owen and C. J. C. Phillips

Processing a feed involves the use of any treatment which alters the composition of that feed by physical, chemical or biological action. In this respect changes in composition do not simply mean improving the nutritional value of the feed but also include a large range of other improvements, such as lengthening the storage life of the feed, detoxification, change in particle size, improving palatability, isolation of specific parts of the feed and reduction in effluent production. Processing a feed may be carried out at any time between harvesting the crop used to produce the feed and delivery of the feed to the livestock.

Processing forages

Straws and other crop residues

In many developing countries, livestock suffer from shortages of feed supply despite the vast amounts of crop byproducts and agroindustrial residues produced that are not completely utilized. Much of these are consumable and utilizable by ruminant animals and can be enriched by different processes, some of which can be carried out by small farmers themselves. In developed countries such feeds are often wasted, e.g. straw burning, when they could replace purchased feed if they could be adequately upgraded. Such materials are characterized by fibrous structure and low protein content. In addition, they are of low digestibility and have low voluntary intake characteristics due to:

(1) poor susceptibility to microbial attack in the rumen caused by the encrustation of the major components (energy stored in the cellulose and hemicellulose in the cell walls) with lignin and/or silica
(2) deficiency in the rumen of essential nutrients for microbial activity, i.e. nitrogen and amino acids as donors of branched chain volatile fatty acids, sulphur, phosphorus, cobalt and possibly water soluble vitamins (Taminga, 1986).

The main treatments for improving the voluntary intake characteristics and nutritive value of crop byproducts and residues are physical, chemical and biological, of which one or more can be applied. However, biological treatment has not so far succeeded on a farm scale. The applicability of the techniques can be widespread, depending on the availability of materials, but generally the feeds are not sufficiently improved to be of value to the high yielding dairy cow, whereas they are of value in improving the performance of less intensive cattle production systems.

Physical treatment

This includes:

(1) milling, grinding and chopping
(2) steam treatment
(3) ionizing radiation.

However, separation and collection of different parts, including the leaves, of crop byproducts during harvesting (preferably without extra costs), can be worthwhile, since leaves are more digestible than stem or chaff (Mowat and Wilton, 1984).

Milling, grinding and chopping

The aim is to increase the surface area available to enzymatic digestion of cellulose by rumen microorganisms and to increase the animal's voluntary intake. The beneficial effect of grinding is offset by fine materials passing through the digestive tract too rapidly, i.e. leaving too short a time for maximum nutrient utilization (Greenhalgh and Wainman, 1972). Also feeding finely ground straw causes a reduction in rumen pH, the ratio of acetic:propionic acid concentrations and the overall concentration of volatile fatty acids, thus favouring a reduction in the synthesis of milk components in the mammary gland of dairy cows. Swan and Clarke (1974) reported that straw ground through a 6 mm screen gave optimal animal performance in terms of diet digestibility, intake and growth rate when compared to straw ground through 3, 10 or 13 mm screens in a beef fattening trial using diets containing 30% ground barley straw.

Steam treatment

High pressure steam treatment of roughages is attractive when energy costs are low, and is characterized by varying degrees of destruction of the hemicellulose fraction, and the production of acetic and other acids, furfural and phenolic derivatives (Walker, 1984). Normally a moisture content of at least 60% is required to prevent charring, and steam pressures of between 7 and 28 kg/cm^2 are applied for periods ranging from a few seconds to several minutes (Klopfenstein and Bolsen, 1971). The reduction in hemicellulose content is directly proportional to the degree of treatment, but at severe treatment levels the digestibility of cell wall contents is reduced, giving rise to an overall reduction in the dry matter (DM) digestibility of the feed. In addition, a non-enzymatic browning reaction at high treatment levels causes an increase in the apparent lignin content and acid-digestible insoluble nitrogen content.

Ionizing radiation

Ionizing radiation of lignocellulosic materials increases their digestibility by decreasing cellulose chain length and increasing the availability of the insoluble carbohydrate components to rumen microorganisms (Lawton et al., 1951; Seaman, Millet and Lawton, 1952).

The process is only likely to be cost effective with material extremely refractory to degradation, such as sunflower hulls or sawdust, because straw-like products respond to less expensive treatment methods (Walker, 1984). The degree of improvement in digestibility is directly related to the dose of irradiation given (McManus *et al.*, 1972).

Chemical treatment

Chemicals used for upgrading crop residues and byproducts are numerous. Selection of a particular chemical depends on its effectiveness in improving digestibility and/or intake, cost of treatment (chemical cost and application), availability and freedom from chemical residues that could be toxic to animals directly or animals and man through faeces and urine polluting soils and water courses. Also the chemical should be non-hazardous to handle by man and non-corrosive to machinery (Owen, Klopfenstein and Urio, 1984). Most chemicals are chosen for their ability to alter the pH outside the range normally tolerated by microorganisms.

Alkali treatment
Sodium hydroxide (NaOH)   Sodium hydroxide is known to be a particularly effective chemical in upgrading crop wastes, and the developments in this technique since Beckmann's original 'wet method' have been reviewed by Jackson (1977), Homb (1984) and Wilkinson (1984). Since the pioneering work on cereal straws the technique has been developed for use on other crop residues such as rice hulls (Hassona, 1986), straw (Abou-Raya *et al.*, 1983), soyabean straw and peanut hulls (Sherif *et al.*, 1985).

Calcium hydroxide (Ca(OH)$_2$)   Calcium hydroxide has been used for upgrading fibrous byproducts instead of sodium hydroxide as it is less expensive and safer to handle and the calcium residue presents no problems for animals. However, it is weaker than sodium hydroxide and needs a long time to react on crop residues (at least 7 days) depending on the ambient temperature. Furthermore, poor solubilities of calcium hydroxide represent a considerable disadvantage to its use and render it less effective. Moisture content of the crop residue is an important factor, since it affects the keeping quality when treated with calcium hydroxide. Residues treated at 40–45% moisture tend to develop moulds, although conditions for delignification are optimum (Bass, Parkins and Fishwick, 1982). However, at 60% moisture delignification is not complete and moulding does not occur, but heating the treated residues with calcium hydroxide at 60% moisture improves the reaction.
    A soaking method was used by Abou-Raya *et al.* (1984) with some modifications. It was carried out by soaking two-thirds of the residues in a solution of calcium hydroxide for 24 h. The remaining part of the residues (one-third) was then added, mixed thoroughly to absorb the drainage and kept for 24 h. The treated residues were spread out to dry and self-neutralize for 7 days. This method resulted in an increase in the organic matter (OM) digestibility of maize stalk by 17% and rice straw by 8%. Spraying different roughages with solutions of calcium hydroxide at different concentrations (2, 4 and 6%) and keeping the treated material in plastic bags for 48 h to react (Attia, 1985) gave increases in the OM digestibility of the treated roughages of 5, 6 and 19 percentage units respectively. However, Abou-Hussein *et al.* (1982) obtained negative results when maize stalks and bean

straw were boiled in a solution of 1% calcium hydroxide and attributed this to changes in the structure of the cells due to boiling.

To overcome the inefficiency of calcium hydroxide for delignification of crop residues the use of a mixture of calcium and sodium hydroxides has considerable potential as well as in increasing voluntary intake and minimizing the sodium content of the treated material. Owen *et al.* (1982) found that when rice straw was treated with a mixture of sodium and calcium hydroxides (45 and 15 g/kg respectively), the dry matter (DM) intake of lambs was 1331 g/day compared with 1227, 1006 and 868 g/day for those fed straw treated with sodium hydroxide, calcium hydroxide, or untreated straw, respectively.

Potassium hydroxide (KOH)   Potassium hydroxide is another alkali which has been used for upgrading crop residues and byproducts. It is as effective as sodium hydroxide in delignification of residues and has the advantage of not increasing the

Figure 6.1 Ammonia treatment of polythene covered straw tunnel

sodium content of the treated residues. However, potassium hydroxide is expensive in the pure form. Wood ash is a crude form of potassium hydroxide, thus in areas where this ash is readily obtainable crop residues could be soaked in solution containing 50 g ash/kg residues (Bergener, 1981).

Ammonia   The use of ammonia either in anhydrous or in aqueous form is popular in many countries especially where protein supplements are scarce and/or expensive. Treatment of roughages by ammonia is reviewed by Sundstol and Coxworth (1984). The response to ammonia has been found to be highly dependent on temperature (Borhami and Sundstol, 1982), straw variety and moisture content (Horn *et al.,* 1983). Straw is usually treated in either a polythene covered stack or tunnel (Figure 6.1) using aqueous ammonia or an oven using anhydrous ammonia.

Urea   A safer alternative to ammonia is urea which is available in most areas as a fertilizer and precursor of ammonia. Urea–molasses mixtures are also available to be mixed with crop residues before feeding to animals or to be used as a solid lick. In the latter case, improved feeding value will be obtained as a result of maximized rumen flora activity.

The use of urea as a precursor of ammonia is recommended for developing countries for its simplicity and safety in application, availability in local markets at cheap prices and preservative properties. In Thailand, the results obtained by Wanapat (1986) when rice straw treated with urea at different concentrations (3 and 5%) either wet or dry was fed to cattle showed that the straw treated with 5% urea had increased DM digestibility and voluntary intake, especially the wet straw. However, although DM digestibility increases with increasing levels of urea, voluntary intake is reduced at high levels (Ibrahim, 1986). In temperate zones results are expected to be poorest when treatment occurs during the cool months of autumn and winter. The reduction in the efficiency of the treatment is because at low ambient temperatures (20°C), some of the CO_2 and NH_3 released by the breakdown of urea subsequently react to produce ammonium carbonate which is a stable compound (Mason and Owen, 1986).

In Egypt, Barker *et al.* (1987) described a system where a mixture of molasses and urea (91.4% molasses, 2.5% urea, 1.1% minerals and 5% water) is distributed to small holders to be mixed with the straw. This mixture proved to be successful in sustaining the milk yield of cows and rations were around 16% cheaper than the berseem (*Trifolium alexandrinum*) hay.

Acid treatment
The main acids tested for their ability to upgrade roughages are sulphuric acid (Holzer, Levy and Folman, 1978; Williams *et al.,* 1979; Fahmy and Ørskov, 1984; Hassona, 1986), hydrochloric acid and chlorine (Arndt *et al.,* 1980; Turner *et al.,* 1985), nitric acid (Arndt *et al.,* 1980) and formic, orthophosphoric or propionic acids (Owen, Perry and Rees, 1977).

Although volatile fatty acids and formic acids are relatively safe to use on a farm scale, they are not effective in upgrading crop residues. On the other hand, sulphuric acid is hazardous to handle under farm conditions, particularly in the developing countries, but is effective as an upgrading agent (Hassona, 1986). Fahmy and Ørskov (1984) increased the DM digestibility of straw by treating with H_2SO_4 at concentrations of 20–60 g H_2SO_4/kg straw DM, but the low pH of the treated products (2.2–4.4) could give intake problems in practice.

Biological treatment

The use of biological treatment (BT) for upgrading lignocellulosic materials (wood and crop residues) is potentially safer and cheaper than chemical and physical treatments, but the contamination of the treated materials during the BT process with unwanted microorganisms may be a disadvantage.

The biological treatment of lignocellulosic materials is based on the use of certain microorganisms that are very efficient in lignin metabolism but with low degradation rates of cellulose and hemicellulose. In addition, these organisms should preferably be able to:

(1) yield protein with the high proportion of essential amino acids required by ruminants
(2) convert cheap nitrogen sources (e.g. urea) into their biomass
(3) grow rapidly in the substrate (crop residues) and preferably in non-sterile conditions
(4) produce a product free of toxins or disease causing agents (Flegel and Meevootisom, 1986).

Crop residues may be pretreated with heat or chemicals, especially if bacteria or yeast are used to produce single cell protein feeds for monogastric animals (Zadrazil, 1984). In this case, the costs of pretreatment should be considered in evaluating the BT.

Different species of fungi have been used for BT of crop residues (Ibrahim and Pearce, 1980; Langar, Seghal and Garcha, 1980; Streeter *et al.*, 1982). The cultivation of edible fungi (*Pleurotus spp., Volvariella rolvacae, Strophania rugoso annulata*) has also been used to convert crop residues into human and animal foods (Kirk, 1983). Most of these studies have been carried out on a small scale in the laboratory. However, the University of Waterloo has introduced a large-scale process for upgrading crop residues using BT (Moo-Young, Chahal and Stickney, 1981; Moo-Young, MacDonald and Ling, 1981). This process has been patented and successfully used in Canada. The residues have to be alkali treated at high temperature, then neutralized and cooled and nitrogen salts are added. Inoculation with the fungus *Chaetomium cellulolyticum* follows and after controlled fermentation for 4 h the protein content of the substrate reaches 12%. This process is effective, but capitally intensive and still under study to overcome two main problems:

(1) the necessity for pretreatment (with heat and/or chemical)
(2) contamination with unwanted microorganisms.

The first could be overcome by adding some soluble carbohydrate (glucose, molasses) to the medium. The second could be solved if the pH of the medium can be kept around 3 without detrimental effect on the growth of fungus.

Poultry manure and litter

Poultry manure from layer cages or broiler litter is a good source of nitrogen and a low price protein replacer which could be processed in the rations of ruminants to reduce feeding costs and environmental pollution. Broiler litter is high in crude protein (15–25%) and ash (15%; mostly calcium and phosphorous) and contains substantial amounts of utilizable energy for ruminants (Fontenot *et al.*, 1971). About half of the crude protein seems to be made up of true protein which is high in glycine and relatively low in arginine, lysine, methionine and cystine (Bhattacharya and Fontenot, 1966). However, the high moisture content (60 and 30% in poultry manure and broiler litter, respectively) has to be reduced and unwanted microorganisms eliminated before inclusion in animal rations. Drying and ensiling either alone or with green or dried roughages may result in dry final products which have no odour problems and are almost sterile so far as salmonella and coliform species are concerned. The use of solar energy as a low cost on-farm drying technique helps in drying and destroying salmonella species (destroyed at a temperature of 68°C for a period of 30 minutes).

Ensiling poultry manure and litter is a simple, safe and cheap process which not only prevents crude protein losses, but also converts part of the available NPN (non-protein nitrogen) into true protein (Muller, 1982). Further, several salmonella serotypes have been reported not to survive in bovine manure ensiled for 3–4 days (McCaskey and Anthony, 1975), since acids produced (mainly lactic acid) during ensiling reduce the pH to 4.0–4.5 (or lower) which is sufficient to kill salmonellae, especially when the ensiling temperature is greater than 25°C.

Good feeding values and acceptable characteristics were reported when poultry manure or litter was ensiled with chopped maize (Gouet and Girardeau, 1982), maize grain (Grotheer et al., 1980), molasses (Muller, 1982), corn cobs and stalks (Vintila et al., 1980).

El Hosseiny (1984) found that the inclusion of 20% poultry litter resulted in the production of good silage in terms of pH and lactic acid content. Voluntary intake of lambs was not affected by treatment and the silage was readily acceptable when ensiling corn stalks with broiler litter at different rates (0–30%).

Silages

Changes in feeding practices have been enforced on the livestock industry in recent years as a result of over-production of meat and dairy products. The imposition of production restraints, such as milk quotas in the European Community, have necessitated reducing feed costs by making greater, and more efficient use of on-farm resources. This need has been reflected in more forage being conserved, particularly in the form of silage (Table 6.1).

Attaining higher levels of production from forage sources requires conservation at optimum quality. This has led to crops being ensiled earlier in the season, more aftermath cuts being ensiled extending into the autumn, and wilting being reduced to a minimum to avoid losses. There have also been important developments in ensiling methodology notably the proliferation of biological additives, incorporation of concentrates in silage and increasing use of bale techniques.

Table 6.1 Conserved forage production, England and Wales
(million tonnes dry matter)

	Silage	Hay	Silage (% of total)
1980	3.75	4.69	44
1985	5.88	3.39	63
1987	6.73	3.14	68

MAFF, 1988; WOAD, 1988.

Recent developments in ensiling methodology

Wilting

There has been an active argument in recent years on the merits of wilting prior to ensiling, particularly with regard to its effect on feed quality and animal performance. Wilting has obvious disadvantages in increasing weather dependence in silage making, particularly with the high rainfall that is characteristic of many livestock areas.

A comparison of unwilted and wilted silages in a series of experiments conducted throughout Europe has been described by Zimmer and Wilkins (1984). The survey concluded that dry matter intake was increased by wilting by about 4% in dairy cows and 9% in fattening cattle. These higher intakes were not, however, translated into improved production partly because wilting reduced DM digestibility by 2–3% units. With dairy cows, milk yield was decreased by 2–3% and weight gain by 7% when feeding wilted compared with unwilted silages. Liveweight gain of growing cattle was reduced by just over 4% as a result of wilting. It was calculated that production per hectare from wilted silage would be 91–95% of that from unwilted silage. It was noted that the performance from wilted silages was markedly lower if field wilting was prolonged or the crop heavily contaminated with soil.

Other studies have shown clearer evidence of improved performance from unwilted silage. For example, Gordon (1986) found that dairy cows consumed 16% less unwilted silage but produced 10% more milk than from wilted silage. The overall effect was a 12% greater milk output from the unwilted silage.

While the balance of evidence may indicate some nutritional advantage in not wilting, the production of increased amounts of effluent from silage pits has become a major environmental problem in many areas. Effluent is characterized by an extremely high demand for biological oxygen and is therefore a potent pollutant which must be prevented from entering watercourses. Retention within the silo or effective collection and dispersal are therefore essential. It has been shown (Steen, 1986) that cattle can be fed on fresh or well preserved effluent without apparent ill health. Effluent was consumed in preference to water and increased DM intake by 10% without detriment to silage intake.

Incorporation of dried concentrates

Silage is commonly fed with concentrates during winter feeding and there are potential advantages to mixing the dried concentrate with the forage prior to ensiling. These include:

(1) the retention of effluent within the clamp thus retaining more sugars for fermentation and reducing the possibility of pollution
(2) the production of a complete mixed diet with possible nutritional and labour saving benefits.

Several studies have shown that incorporation of cereals at ensiling improves animal performance compared to conventional feeding of cereals as discrete supplements to grass silage. Nicholson and MacLeod (1966) showed that the incorporation of 7.5% of wilted barley at ensiling increased silage intake and liveweight gain in cattle. In a series of experiments (Sporndly, Burstedt and Lingvall, 1982; Sporndly, 1986) it has been shown that the inclusion of 15% rolled oats at ensiling resulted in higher silage intakes and milk yield. Other work with beef cattle has given conflicting results in different experiments (Stewart, 1966, 1967).

Recent experiments at the Welsh Plant Breeding Station using rolled barley and dried sugar beet pulp have confirmed the potential advantages of concentrate incorporation during ensiling. Effluent production was halved by the addition of 50 kg rolled barley/tonne of grass in the silo in comparison with untreated or formic acid treated grass. Silage fermentation was improved by the addition of barley and comparable to that from formic acid treatment, and the DM and metabolizable energy content were significantly higher for the barley added silage (Table 6.2).

Table 6.2 Effect of barley incorporation on silage quality

Treatment	Dry matter %	Ammonia N (% of total N)	ME (MJ/kg DM)
Untreated	16	11	10.1
Barley added	19	8	10.6
Formic acid treated	16	5	10.3

ME: metabolizable energy.

Jones and Jones, 1988.

In feeding experiments with housed beef cattle silage DM intakes were significantly higher than for the untreated and formic acid treated silages, these being fed with a daily supplement of barley equivalent to that contained in the barley added silages. Liveweight gains averaged 1.00 kg/head per day for the barley added silage compared with 0.96 kg for the formic acid silage and 0.82 kg for the untreated control. In other experiments the inclusion of dried sugar beet pulp (50 kg/tonne grass) has been shown to produce silage of comparable fermentation quality to that from formic acid treatment. Liveweight gains in beef cattle averaged 10% higher for the sugar beet incorporated silages than for the same grass silage fed with an equivalent amount of sugar beet (Jones and Jones, 1988).

A further development is the incorporation of dried concentrates in baled silage using a commercially developed applicator (Jones, 1988). The inclusion of concentrate in bales has obvious advantages in certain feeding situations as it avoids the necessity for daily feeding.

Biological additives
In recent years considerable interest has centred on inoculants, enzymes, or mixtures of both as silage additives. Their obvious advantages is their non-corrosive nature when compared with conventional acid or acid mixtures.

Inoculants are intended to increase the content of lactic acid bacteria available for fermentation and contain one or more species of lactic acid bacteria, often with the addition of nutrients. Enzymes are generally crude cellulases able to digest cell wall cellulose and hemicellulose during ensiling thereby increasing the available sugar supply. The use of biological additives has been reviewed recently by McDonald (1981), Woolford (1984) and Seale (1987).

While many laboratory scale studies have shown inoculants and enzymes to improve fermentation there are few reports from large scale silos on animal performance. A recent study by Haigh, Appleton and Clench (1987) concluded that, while formic acid addition improved fermentation and liveweight gain, the commercial inoculants tested provided little benefit compared with the untreated silage. Owen (1986), however, showed an inoculant to improve fermentation and to improve the intake and liveweight gain of cattle by some 10%, while Gordon (1987) found an inoculant to increase intake and milk yield in dairy cows.

The present evidence for the efficacy of biological additives is, therefore, not unequivocal and probably reflects variation in the numbers of naturally occurring lactic acid bacteria present, soluble carbohydrate levels and other factors.

Processing cereals and other concentrate feeds

Cereals as cattle feeds are mainly processed to increase nutrient availability, add to the nutrient content, improve keeping quality and detoxify undesirable ingredients. Since they cannot utilize whole grains as effectively, the processing of cereal grains is more important for cattle than for sheep or goats. At its simplest cereal grain processing involves drying to improve keeping quality and this technique, together with rolling, grinding and pelleting, is an established practice. More recently, physical processing techniques such as extruding, steam flaking and jet-sploding (rapid heating to very high temperatures) have been investigated as more efficient methods of increasing nutrient availability than rolling or grinding. In addition, a number of chemical agents such as sodium hydroxide or ammonia are now used to preserve moist grain. The nutritional value of fats, proteins and amino acids extracted from plant materials may be enhanced by protecting them from rumen microbial degradation so that they pass directly into the abomasum and are absorbed into the bloodstream from the intestine.

Processing concentrates for upgrading and detoxification purposes

The main cereals fed to cattle include wheat, barley, maize and sorghum which have starch and protein degradation rates in the order wheat > barley > maize > sorghum (Theurer, 1986; Herrera-Saldana, Huber and Poore, 1988). The magnitude of improvement in starch digestibility with processing is inversely related to the extent of starch degradation in the unprocessed grain (Theurer, 1986). Modern methods of cereal processing involve combinations of moisture, heat and pressure application. A process that aims to apply these three processing methods at the optimum level is extrusion.

Extrusion of cereals is currently used to produce mainly human, pet and fish foods and also to increase the digestion by yeast in alcoholic beverage production and baking (Briggs et al., 1986; Statham, 1987). The increase in starch digestibility occurs through a physiochemical change in the starch granules called gelatinization. Both temperature and moisture levels have marked effects on gelatinization. Asp (1987) and Woodruff and Webber (1933) have shown that hard and soft wheats have different temperature requirements for gelatinization to occur. Gelatinization increases starch digestibility in the rumen, although postruminal digestion of starch is reduced. As a result increased levels of rumen degradable nitrogen may need to be given if an increased microbial population is to be sustained in the rumen. Research in the USA with a combination of extruded maize and urea showed increased milk yield compared with unextruded maize (Helmer, Bartley and Deyoe, 1970), and extrusion of sorghum grains has also been shown to increase milk yields (Collenbrander *et al.*, 1967). No effect of extrusion of barley or wheat on milk production has been found, although extrusion of wheat together with a urea supplement was found to increase milk protein content (Hecheimi and Phillips, 1988).

A further effect of starch gelatinization is to reduce the density of the cereal and increase its swelling capacity. Preliminary research has shown extruded wheat to be more effective than ground wheat when used as an additive to silage to absorb effluent (Phillips, unpublished data).

Extrusion and jet-sploding can have both beneficial and detrimental effects on proteins in concentrates. At high temperatures the Maillard reaction between sugar aldehyde groups and free amino acid groups can render the latter unavailable, but more moderate heat will reduce protein degradability in the rumen by creating cross linkages both within and between peptide chains and with carbohydrates, while not affecting total tract amino acid digestibility (Mielke and Schingoethe, 1981; Deacon, De Boer and Kennelly, 1988). No significant effects of extruding high protein concentrates, e.g. soyabeans, on milk production were found by Mielke and Schingoethe (1981), although the activity of the potentially toxic trypsin inhibitor in soyabeans is largely eliminated. The extrusion process is also known greatly to reduce the levels of some vitamins, depending on the method used (Schlude, 1987).

Processing concentrates for preservation and rumen protection

The use of alkalis such as sodium hydroxide to preserve high moisture grain is well established and it has been found that both maize and sorghum, but not barley, are more efficiently utilized when preserved by this method than by drying (Kennelly, Dalton and Ha, 1988). Ammonium hydroxide is a useful alternative to sodium hydroxide where supplementary rumen degradable nitrogen is required or where excess sodium intake is detrimental; it has been shown effectively to preserve high moisture maize (Soderholm *et al.*, 1988).

Recently sodium hydroxide has been used to reduce the protein degradability of high protein feeds such as soyabeans (Mir *et al.*, 1984; Bowman *et al.*, 1988). This method is more reliable than treatment with heat (Thomas, Trenkle and Burroughs, 1979), formaldehyde (Sharma and Ingalls, 1974; Mir *et al.*, 1984) or tannins (Nishimuta, Ely and Boling, 1974). These latter treatments frequently result in underprotection or alternatively overprotection so that the protein is indigestible. The use of blood to protect high protein feeds from rumen degradation has been successfully demonstrated (Ørskov, Mills and Robinson, 1980) and probably works because the blood which is not easily degraded in the rumen, provides a protective coating. Sodium hydroxide has also been recently used to treat soya flour to produce a vegetable-based milk replacer for calves (Kelly and Ramsey, 1988). Although this milk replacer will not as yet produce growth rates as high as replacers based on skimmed milk powder, further research on heat and alkali treatment of soya flour will probably achieve this aim.

The nutritional benefits of supplementing the diets of high-producing ruminants with the essential amino acids that are in deficit (particularly methionine and lysine) and with fats, has led to considerable efforts to achieve reliable methods of rumen protection so that they can be absorbed directly. Two forms of protected methionine have been produced – encapsulated methionine and methionine analogues or polymeric compounds (Papas *et al.*, 1984). In early preparations of encapsulated methionine, the amino acid was embedded in a matrix of protein or fat treated with formaldehyde, but this was found to be only partially stable in the rumen and postruminal release of amino acid was poor (Chalupa, 1975). Later, polymeric coatings that are stable at the higher rumen pH but dissociate at the lower abomasol pH, were successfully developed (Papas et al., 1984). Recently, methionine has been successfully compounded with the calcium salts of fatty acid (Casper et al., 1987) which has produced elevated blood methionine levels and increased milk protein concentrations.

Methionine hydroxy analogue has been found to be biologically active in monogastrics but unstable in the rumen and has produced inconsistent production responses (Chalupa, 1975).

High concentrations of fat, particularly unsaturated fat in the rumen, impair the microbial digestion of long fibre and depress appetite. The first protected lipids were produced by Scott, Cook and Mills (1971), who encapsulated lipid in a layer of formaldehyde-treated casein, and then spray dried the resulting emulsion. The high cost of casein and spray drying has limited the development of this concept. Scott et al. (1972) have also protected whole oil seeds by removal of the seed husks, solubilization of the native protein with alkali, emulsification, formaldehyde treatment and drying.

Recently it has been found that feeding preformed calcium salts of long chain fatty acids or a mixture of highly soluble calcium salts with free fatty acids, will minimize the adverse effects of supplementary fat on the rumen (Gummer, 1988), although some depression in forage intake when feeding high levels of the calcium salt of palm oil has been noted (Phillips, Angold and Statham, 1988). In cases where forage:concentrate ratio and rumen pH are low it may be necessary to feed alkali buffers to restore rumen pH to a normal level and prevent dissociation of the calcium soaps.

Processing and mixing ingredients into complete diets

Advances in the design of feed handling machinery and, in particular, equipment for mixing and dispensing diets has helped the development of the complete diet system for feeding cattle (Owen, 1983). Complete diets can be used as the sole diet for yarded cattle, as a basal allowance supplemented by concentrate feed or as a buffer feed for grazing cattle (NOSCA, 1976). The system has proved suitable for feeding cattle in large units (McGillard, Swisher and James, 1983) and for incorporating a wide range of byproducts into the diet (Kroll and Owen, 1986).

Although various attempts have been made to process complete diets into small packages, e.g. pellets, cubes, nuts or 'briquettes' (Borodulin and Popekhina, 1978), not many have yet led to techniques of wide applicability. In practice, complete diets are therefore based on a mixed blend of ingredients presented in the form of a loose mix of forage and concentrates.

In processing feed ingredients for incorporation into complete diets the prime requirement is to reduce particle size to a level which facilitates the production of a uniform blend that is not easily consumed selectively by the cattle. At the same time, it is important to ensure that particle size is not reduced below the level where undesirable effects are produced on rumen fermentation and lactational performance, e.g. the low fat milk syndrome. Several studies have confirmed that above certain critical particle size levels there seem to be no important effects on

Figure 6.2 A recently developed mixer wagon incorporating a paddle mixing mechanism. (*a*) Mixer wagon; (*b*) interior. (Courtesy of Richard Keenan & Co Ltd)

lactating cattle (Armentano, Pastore and Hoffman, 1988). The critical particle size appears to depend on the forage level in the diet and on the quality of the forage used (Shaver et al., 1988).

Machinery widely used to process the forage before mixing is based usually on the harvesting mechanism, particularly forage harvesters. The use of precision chop equipment as against coarser chop machines has not shown much advantage in terms of fermentation and animal performance (Castle, Retter and Watson, 1979; Gordon and Unsworth, 1986), but precision chopped material is easier to mix, particularly in screw auger type mixer wagons, and is less likely to cause problems of feed ingredient selectivity by the cattle. Dry forages, including straws and hulls from a wide range of material, are often harvested and stored in bales and mostly require some pretreatment, usually chopping, to achieve the desired particle size for mixing.

Non-forage materials such as cereals, also generally require processing before feeding to cattle to avoid losses in digestibility (Broadbent, 1976). Roots such as swedes and fodder beet also usually require chopping into small pieces for successful incorporation into mixed diets.

The key feature of the processing of complete diets in practice has been the development of mixer wagons. These mobile units have the capacity to mix 2–7

Figure 6.3 Mixer wagon with 15 m^3 capacity in operation on an Israeli farm. (Courtesy of Lachish Industries Ltd)

tonnes of complete diet loaded at the place where ingredients are stored and to dispense the diet at the cattle feed face (Figure 6.2). The design varies but most machines are based on the opposing screw auger mixing principle or on a paddle system. Machines based on the latter system have proved to be simple and robust and are popular in the UK. Some heavy duty machines used in the USA and in Israel are capable of mixing diets including long fibrous materials such as whole bales of hay and straw without prechopping (Figure 6.3). Lehmer (1981) has assessed the use of various types of mixer wagons in use in Germany.

Evidence is available that cattle consume the lower quality fraction of forage and other ingredients more readily when offered *ad libitum* in a complete diet than when offered the same forage separately (Bines, 1985). There is also some evidence that voluntary feed intake of some mixed diets and the fat content of the milk of lactating animals is higher than when the same ingredients are given separately, although it is difficult to ensure a strictly comparable composition of dietary intake (Phipps *et al.*, 1984; Noceck, Steele and Braund, 1986). If intake is elevated then some of the rise may be due to the effect of the processing and abrasion that accompany the mixing process.

One problem in the processing of complete diets relates to the moisture content of the ingredients and of the mix as a whole. Moisture content can be important for the storage life of the mixture at the feeding place and can have certain animal effects. Studies on the storage life of silage and of complete diets have shown that moisture content and temperature interact in determining the rate of deterioration of the material when loosely dispensed and exposed to air. At air temperatures below 10°C storage life of material is long enough to provide flexibility for most feeding requirements. At higher temperatures storage life is reduced, particularly for mixtures of wet forage such as silage and cereals (ADAS, 1980). Material of over 40% DM in the total mix is generally suitable for most cattle feeding situations where feed is dispensed to cattle at least once per day.

Moisture content is also important from the point of view of cattle intake and performance. Kroll and Owen (1986) have concluded that high moisture diets do seem to limit intake, although the precise cause seems to be complicated. Water added to diets, i.e. extracellular water does not have the same intake depressing effect as intracellular water contained within the feed ingredient. However, it is not easy to design experiments to test the effect unambiguously. Complete diets with more than 15% extra units of moisture deriving from wet materials like citrus pulp seem to depress intake by about 7% compared with the drier, control diets.

A recent development in the complete diet system is the possibility of creating diets that are complete as they are put in store, usually in clamps. This can be achieved by harvesting materials such as arable cereal based crops as silage at a stage where they form the desired complete diet in themselves or at least with only a small amount of additive. The incorporation of cereals, beet pulp and other materials into silage clamps at harvesting is another aspect of this technique (Jones, 1988; Jones and Jones, 1988).

References

Abou-Hussein, E. R. M., Abou-Raya, A. K., Shalaby, A. S. and Salem, O. A. I. (1982) Nutritional and physiological studies with short time alkali treated roughages. 1. Maize stalks as affected by different methods of treatment. *Sixth International Conference on Animal and Poultry Production*, Zagazig, Egypt (ed. A. S. El-Shobokshy, H. M. Ali, A. M. Aboul-Naga, F. Z. Swindon and A. A. Daader). Zagazig University Press, Zagazig, pp. 25–39

Abou-Raya, A. K., Hussein, E. L., Shalaby, A. S. and Salem, O. A. (1983) Nutritional and physiological studies with short time alkali treated roughages. *The 2nd African Workshop*. Alexandria University, Egypt

Abou-Raya, A. K., Shalaby, A. S., Abdel-Motagally, Z. M. Z., Salem, O. A. I. and Salem, F. M. M. (1984) Nutritional studies with modified Ca(OH)$_2$ method (2/3 soaked and 1/3 moistened) versus water boiling with rice straw and maize stalks. 1. Effect of treatment on chemical composition, feeding value and *in situ* digestion. *First Egyptian-British Conference on Animal and Poultry Production*, Zagazig, Egypt, (ed. I. F. M. Marai, S. M. Abd El-Baki and A. Daader), pp. 113–121

Agricultural Development and Advisory Service (ADAS) (1980) *Complete diet feeding of dairy cows*. Supplementary report on investigations 1978–1979. 38 pp

Armentano, L. E., Pastore, S. E. and Hoffman, P. C. (1988) Particle size reduction of alfalfa silage did not alter nutritional quality of high forage diets for dairy cattle. *Journal of Dairy Science, 71*, 409–413

Arndt, D. L., Richardson, C. R., Albin, R. C. and Sherrod, L. B. (1980) Digestibility of chemically treated cotton by-product and effect on mineral balance, urine volume and pH. *Journal of Animal Science, 51*, 215–223

Asp, N. G. (1987) Nutritional aspects: what happens to the different materials at different temperatures? In *Extrusion Technology for the Food Industry*, (ed. C. O'Connor). Elsevier, London, pp. 16–21

Attia, A. I. (1985) Improving the nutritive value of some feeding stuffs by chemical and/or physical treatment. *M.Sc. Thesis*, University of Zagazig, Egypt

Barker, T. J., Yackout, H., Creek, M. J., Hathout, M. *et al.* (1987) Transfer of feeding systems from the research phase to the farmer. *World Animal Review, 61*, 17–25

Bass, J. M., Parkins, J. J. and Fishwick, G. (1982) The effect of calcium hydroxide treatment on the digestibility of chopped oat straw supplemented with a solution containing urea, calcium, phosphorus, sodium, trace elements and vitamins. *Animal Feed Science and Technology, 7*, 93–100

Bergener, H. (1981) Chemical treatment of straw. *Institute for Scientific Co-operation of Recent German Contributions*. Landhausstrasse 18, 7400 Tubingen, pp. 61–81

Bhattacharya, A. N. and Fontenot, J. P. (1966) Protein and energy value of peanut hull and wood shaving poultry litters. *Journal of Animal Science, 25*, 367–371

Bines, J. A. (1985) Feeding systems and food intake by housed dairy cows. *Proceedings of the Nutrition Society, UK, 44*, 355–362

Borhami, B. E. A. and Sundstol, F. (1982) Studies on ammonia-treated straw. 1. The effects of type and level of ammonia, moisture content and treatment time on the digestibility *in vitro and* enzyme soluble organic matter of oat straw. *Animal Feed Science and Technology, 7*, 45–51

Borodulin, E. N. and Popekhina, P. S. (1978) Effectiveness of briquetted and pelleted feeds in commercial production of milk and meat. *Dairy Science Abstracts, 45*, abstract 42, 5

Bowman, J. M., Grieve, D. G., Buchanan-Smith, J. G. and Macleod, G. K. (1988) Response of dairy cows in early lactation to sodium hydroxide treated soybean meal. *Journal of Dairy Science, 71*, 982–989

Briggs, D. E., Wadeson, A., Statham, R. and Taylor, J. F. (1986) The use of extruded wheat, barley and maize adjustments in mashing. *Journal of the Institute of Brewing, 92*, 468–474

Broadbent, P. J. (1976) The utilisation of whole or bruised barley grain of different moisture contents given to beef cattle as a supplement to a forage-based diet. *Animal Production, 23*, 165–171

Casper, D. P., Schingoethe, D. J., Yang, C. M. J. and Mueller, C. R. (1987) Protected methionine supplementation with extruded blend of soybean and soybean meal for dairy cows. *Journal of Dairy Science, 70*, 321–330

Castle, M. E., Retter, W. C. and Watson, J. N. (1979) Silage and milk production: comparison between grass silage of three different chop lengths. *Grass and Forage Science, 34*, 293–301

Chalupa, W. (1975) Rumen bypass and protection of proteins and amino acids. *Journal of Dairy Science, 58*, 1198–1218

Collenbrander, V. F., Bartley, E. E., Morrill, J. L., Deyoe, C. W. *et al.* (1967) Feed processing II. Effect of expanded grain and finely ground hay on milk composition, yield and rumen metabolism. *Journal of Dairy Science, 50*, 1966–1972

Deacon, M. A., De Boer, G. and Kennelly, J. J. (1988) Influence of jet-sploding and extrusion on ruminal and intestinal disappearance of canola and soybeans. *Journal of Dairy Science, 71*, 745–753

El-Hosseiny, M. A. (1984) Nutritional studies on silage. *PhD Thesis*, University of Zagazig, Egypt

Fahmy, S. T. M. and Ørskov, E. R. (1984) Digestion and utilization of straw. 1. The effect of different chemical treatments on degradability and digestibility of barley straw by sheep. *Animal Production*, **38**, 69–74

Flegel, T. W. and Meevootisom, V. (1986) Biological treatment of straw for animal feed. In *Rice Straw and Related Feeds in Ruminant Rations*, (ed. M. N. H. Ibrahim and J. B. Schiere). Department of Tropical Animal Production, Agricultural University, Wageningen, The Netherlands, pp. 181–191

Fontenot, J. P., Webb, K. E. Jr, Harmon, B. W., Tucken, R. E. *et al.* (1971) Studies of processing, nutritional value and palatibility of broiler litter for ruminants. *Proceedings of International Symposium on Livestock Wastes*, **271**, 301–304

Gordon, F. J. (1986) The effect of system of silage harvesting and feeding on milk production. *Grass and Forage Science*, **41**, 209–219

Gordon, F. J. (1987) An evaluation of an inoculant as an additive for grass silage being offered to dairy cattle. *Proceedings of the Eighth Silage Conference*, Hurley, September 1987, 19–20

Gordon, F. J. and Unsworth, E. F. (1986) The effects of silage harvesting system and supplementation of silage-based diets by protein and methionine hydroxy analogue on the performance of lactating cows. *Grass and Forage Science*, **41**, 1–8

Gouet, P. and Girardeau, J. P. (1982) Fermentation behaviour of maize silage supplemented with poultry litter or poultry excreta. Bulletin Technique, *Centre de Recherches Zootechniques et Veterinaires de Theïx* (cited by Herbage Abstracts, **52**, 1102)

Greenhalgh, J. F. D. and Wainman, F. W. (1972) The nutritive value of processed roughages for fattening cattle and sheep. *Proceedings of British Society of Animal Production*, pp. 61–72

Grotheer, M. D., Cross, D. L., Welter, J. F. and Caldwell, W. J. (1980) Fermentation of caged layer waste with dry feeds. *Journal of Animal Science*, **51**, 32

Gummer, R. R. (1988) Influence of prilled fat and calcium salt of palm oil fatty acids on ruminal fermentation and nutrient digestibility. *Journal of Dairy Science*, **71**, 117–123

Haigh, P. M., Appleton, M. and Clench, S. F. (1987) Effect of commercial inoculant and formic acid + formalin silage additives on silage fermentation and intake and on liveweight change of young cattle. *Grass and Forage Science*, **42**, 405–410

Hassona, E. M. (1986) The utilization of treated feeding stuffs in feeding ruminant animals. *PhD Thesis*, University of Zagazig, Egypt

Hecheimi, K. and Phillips, C. J. C. (1988) *Preliminary report – Concentrate extrusion for dairy cows*. Dairy Research Unit, University of Wales, Bangor, UK

Helmer, L. G., Bartley, E. E. and Deyoe, C. W. (1970) Feed processing VI: Comparison of starch, urea and soya bean meal as protein sources for lactating cows. *Journal of Dairy Science*, **53**, 883–887

Herrera-Saldana, R., Huber, J. T. and Poore, M. H. (1988) *In vitro* and *in situ* dry matter, crude protein and starch degradability of five cereal grains. *Journal of Dairy Science*, **71**, (suppl. 1), 177 (abstract)

Holzer, Z., Levy, D. and Folman, Y. (1978) Chemical processing of wheat straw and cotton by-product for fattening cattle. 2 – Performance of animals receiving material after drying and pelleting. *Animal Production*, **27**, 147–159

Homb, T. (1984) Wet treatment with sodium hydroxide. In *Straw and Other Fibrous By-products as Feed*, (ed. F. Sundstol and E. Owen). Developments in Animal and Veterinary Sciences, 14, Elsevier, pp. 106–126

Horn, G. W., Batchelder, D. G., Manor, G., Streeter, C. L. *et al.* (1983) Ammoniation of wheat straw and native grass hay during baling of large round bales. *Animal Feed, Science and Technology*, **8**, 35–46

Ibrahim, M. N. M. (1986) Efficiency of urea-ammonia treatment. In *Rice Straw and Related Feeds in Ruminant Rations*, (ed. M. N. H. Ibrahim and J. B. Schiere). Department of Tropical Animal Production, Agricultural University, Wageningen, The Netherlands, pp. 171–179

Ibrahim, M. N. M. and Pearce, G. R. (1980) Effects of white rot fungi on the composition and *in vitro* digestibility of crop by-products. *Agricultural Wastes*, **2**, 199–205

Jackson, M. G. (1977) The alkali treatment of straw. *Animal Feed Science and Technology*, **2**, 105–130

Jones, R. (1988) Incorporation of dried additives in big bale silage. *British Grassland Society, Regional Meeting*, Aberystwyth, September 1988

Jones, R. and Jones, D. I. H. (1988) Effect of incorporating molassed sugar beet in grass silage. *British Sugar Technical Conference*, Eastbourne 1988, 32 pp

Kelly, F. M. and Ramsey, H. A. (1988) Effect of alkali treatment on the value of soy flour for preruminant calves. *Journal of Dairy Science,* **71**, (suppl. 1), 125 (abstract)

Kennelly, J. J., Dalton, D. L. and Ha, J. K. (1988) Digestion and utilization of high moisture barley by dairy cows. *Journal of Dairy Science,* **71**, 1259–1266

Kirk, T. K. (1983) Degradation and conversion of lignocellulosics. In *The Filamentous Fungi*, (ed. J. E. Smith, D. R. Berry and B. Kristiansen). Vol. 4, Edward Arnold, London

Klopfenstein, T. J. and Bolsen, K. K. (1971) High temperature pressure treated crop residues. *Journal of Animal Science,* **33**, 290 (abstract)

Kroll, O. and Owen, J. B. (1986) Moisture content and forage source in complete diets for the dairy cow. *Research and Development in Agriculture,* **3**, 47–54

Langar, P. N., Seghal, J. P. and Garcha, H. S. (1980) Chemical changes in wheat and paddy straw after fungal cultivation. *Indian Journal of Animal Sciences,* **50**, 942–946

Lawton, E. J., Bellamy, W. D., Hungate, R. E., Bryant, M. P. *et al.* (1951) Some effects of high velocity electrons on wood. *Science,* **113**, 380–382

Lehmer, M. (1981) Production of basic and concentrate feed mixes in feeder wagons and their use in dairy cattle production. *Dairy Science Abstracts,* **46**, Abs 3379, 386

McCaskey, T. A. and Anthony, W. B. (1975) Health aspects of feeding animal waste conserved in silage. *Proceedings of 3rd International Symposium on Livestock Wastes*, University of Illinois, USA, pp. 230–233

McDonald, P. (1981) *The Biochemistry of Silage*, Wiley and Sons, Chichester

McGillard, M. L., Swisher, J. M. and James, R. E. (1983) Grouping lactating cows by nutritional requirements for feeding. *Journal of Dairy Science,* **66**, 1084–1093

McManus, W. R., Manta, L., McFarlane, J. D. and Gray, A. C. (1972) The effects of diet supplements and gamma irradiation on dissimilation of low-quality roughages by ruminants. 2. Effects of gamma irradiation and urea supplementation on dissimilation in the rumen. *Journal of Agricultural Science, Cambridge,* **79**, 41–53

Mason, V. C. and Owen, E. (1986) Urea versus ammonia for upgrading graminaceous materials. *Proceedings of a Workshop held at the University of Alexandria, Egypt*, October, 1985 (ARNAB)

Mielke, C. D. and Schingoethe, D. J. (1981) Heat treated soybeans for lactating cows. *Journal of Dairy Science,* **64**, 1579–1585

Ministry of Agriculture, Fisheries and Food (MAFF) (1988) *Agricultural Census for UK and England.* Ministry of Agriculture, Fisheries and Food, London

Mir, Z., Macleod, G. K., Buchanan-Smith, J. G., Grieve, D. G. *et al.* (1984) Methods for protecting soybean and canola proteins from degradation in the rumen. *Canadian Journal of Animal Science,* **64**, 853–865

Moo-Young, M., Chahal, D. S. and Stickney, B. (1981) Pollution control of swine manure and straw by conversion to Chaetomium cellulolyticum SCP feed. *Biotechnical Bioengineer,* **23**, 2407–2415

Moo-Young, M., MacDonald, G. and Ling, A. (1981) Improved economics of the Waterloo SCP process by increased growth rates. *Biotechniqual Lettres,* **3**, 149–152

Mowat, D. N. and Wilton, B. (1984) Whole crop harvesting, separation and utilization. In *Straw and Other Fibrous By-products as Feed*, (ed. F. Sundstol and E. Owen). Developments in Animal and Veterinary Sciences, 14, Elsevier, pp. 293–304

Muller, Z. (1982) Feed from animal waste: Feeding manual. FAO, *Animal Production and Health Bulletin*, 28

Nicholson, J. W. G. and Macleod, L. B. (1966) Effect of form of nitrogen fertilizer, a preservative and a supplement on the value of high moisture grass silage. *Canadian Journal of Animal Science,* **46**, 71–82

Nishimuta, J. F., Ely, D. G. and Boling, J. A. (1974) Ruminal bypass of dietary soybean protein treated with heat, formalin and tannic acid. *Journal of Animal Science,* **39**, 952–957

Noceck, J. E., Steele, R. L. and Braund, D. G. (1986) Performance of dairy cows fed forage and grain separately versus a total mixed ration. *Journal of Dairy Science,* **69**, 2140–2147

North of Scotland College of Agriculture (NOSCA) (1976) Buffer feeding of dairy cows at grass. *Research Investigation and Field Trials*, 1974–75, p. 18

Ørskov, E. R., Mills, C. R. and Robinson, J. J. (1980) The use of whole blood for the protection of organic materials from degradation in the rumen. *Proceedings of the Nutrition Society*, **39**, 60A

Owen, E., Klopfenstein, T. J., Britton, R. A., Rump, K. *et al.* (1982) Treatment of wheat straw with different alkalies. *American Society of Animal Science*, **806** (abstract)

Owen, E., Klopfenstein, T. and Urio, N. A. (1984) Treatment with other chemicals. In *Straw and Other Fibrous By-products as Feed*, (ed. F. Sundstol and E. Owen). Developments in Animal and Veterinary Sciences, 14, Elsevier, pp. 248–275

Owen, E., Perry, F. G. and Rees, P. (1977) The effect of various chemicals on the *in vitro* digestibility of wheat straw under ensilage conditions. *Report on research study by University of Reading and BP Nutrition (UK) Ltd*

Owen, J. B. (1983) *Cattle Feeding*, Farming Press Ltd, Ipswich, p. 170

Owen, T. R. (1986) In *Developments in Silage 1986*. Chalcombe Publications, Marlow, pp. 34–44

Papas, A. M., Vicini, J. L., Clarke, J. H. and Pierce-Sandher, S. (1984) Effect of rumen protected methionine on plasma amino acids and production by dairy cows. *Journal of Nutrition*, **114**, 2221–2226

Phillips, C. J. C., Angold, M. G. and Statham, R. (1988) The effect of feeding protected fat and methionine supplements and cereal extrusion on the production and milk characteristics of dairy cows. *Journal of Dairy Science*, (in press)

Phipps, R. H., Bines, J. A., Fulford, R. J. and Weller, R. F. (1984) Complete diets for dairy cows: a comparison between complete diets and separate ingredients. *Journal of Agricultural Science, Cambridge*, **103**, 171–180

Schlude, M. (1987) The stability of vitamins in extrusion cooking. In *Extrusion Technology for the Food Industry*, (ed. C. O'Connor). Elsevier, London, pp. 22–34

Scott, J. W., Cook, L. J. and Mills, S. C. (1971) Protection of dietary polyunsaturated fatty acids against microbial hydrogenation in ruminants. *Journal of the American Oil and Chemical Society*, **48**, 358–364

Scott, J. W., Bready, P. J., Royal, A. S. and Cook, L. J. (1972) Oil seed supplement for the production of polyunsaturated ruminant milk fat. *Search*, **3**, 170–171

Seale, D. R. (1987) Bacteria and enzymes as products to improve silage preservation. In *Developments in Silage 1987*, Chalcombe Publications, Marlow, pp. 47–61

Seaman, J. F., Millet, M. A. and Lawton, E. J. (1952) Effect of high energy cathode rays on cellulose. *Industrial Engineer Chemistry*, **44**, 2848–2852

Sharma, H. R. and Ingalls, J. R. (1974) Effect of treating rapeseed meal and casein with formaldehyde on apparent digestibility and amino acid composition of rumen digesta and bacteria. *Canadian Journal of Animal Science*, **54**, 157–167

Shaver, R. D., Nytes, A. J., Satter, L. D. and Jorgensen, N. A. (1988) Influence of feed intake, forage physical form and forage fiber content on particle size of masticated forage, ruminal digesta and feces of dairy cows. *Journal of Dairy Science*, **71**, 1566–1572

Sherif, S. Y., El-Shobokshy, A. S., Abdel-Rahman, G. A., Abdel-Baki, S. and Attia, A. I. (1985) Improving the nutritive value of some poor quality roughages by physical and chemical treatments. 1. Leguminous by-products. *Annals of Agricultural Sciences, Moshtohor*, **23**, 123–134

Soderholm, C. G., Otterby, D. E., Linn, J. G., Hansen, W. P. *et al.* (1988) Addition of ammonia and urea plus molasses to high moisture snapped ear corn at ensiling. *Journal of Dairy Science*, **71**, 712–721

Sporndly, R. (1986) Ensiling of blended grass and grain and its utilization by dairy cows. *Report 155*, Swedish University of Agriculture Sciences, Uppsala 1986

Sporndly, R., Burstedt, E. and Lingvall, P. (1982) Grass grain silage for lactating dairy cows. *Proceedings of the 9th European Grassland Federation*. Occasional Symposium of the British Grassland Society, no. 14, Reading 1982

Statham, R. (1987) Case study on pet foods. In *Extrusion Technology for the Feed Industry*, (ed. C. O'Connor). Elsevier, London, pp. 90–95

Steen, R. W. J. (1986) An evaluation of effluent from grass silages as a feed for beef cattle offered silage based diets. *Grass and Forage Science*, **41**, 39–46

Stewart, T. A. (1966) An evaluation of barley meal as an additive for autumn made silage. *Annual Report*, Ministry of Agriculture for Northern Ireland, 1967, p. 108

Stewart, T. A. (1967) An evaluation of barley meal as an additive for autumn made silage. *Annual Report*, Ministry of Agriculture for Northern Ireland, 1967, p. 120

Streeter, C. L., Conway, K. E., Horn, G. W. and Mider, T. L. (1982) Nutritional evaluation of wheat straw incubated with the edible mushroom, *Pleurotus ostreatus. Journal of Animal Science,* **54,** 183–188

Sundstol, F. and Coxworth, E. M. (1984) Ammonia treatments. In *Straw and other Fibrous By-products as Feed,* (ed. F. Sundstol and E. Owen). Developments in Animal and Veterinary Science, 14, Elsevier, pp. 196–247

Swan, H. and Clarke, V. J. (1974) The use of processed straw in rations for ruminants. In *University of Nottingham Nutrition Conference for Feed Manufacturers-8.* Butterworths, London

Taminga, S. (1986) Prospects for supplementation of crop residues in tropical countries. In *Rice Straw and Related Feeds in Ruminant Rations*, (ed. M. N. M. Ibrahim and J. B. Schiere). Department of Tropical Animal Production, Agricultural University, Wageningen, The Netherlands, pp. 208–217

Theurer, C. B. (1986) Grain processing effects on starch utilization by ruminants. *Journal of Animal Science,* **63,** 1649–1662

Thomas, E., Trenkle, A. and Burroughs, W. (1979) Evaluation of protective agents applied to soybean meal and fed to cattle 1. Laboratory measurements. *Journal of Animal Science,* **49,** 1337–1345

Turner, N. D., Schelling, G. T., May, B. J., Greene, L. W. *et al.* (1985) *In vitro* and *in situ* digestibility of various hays after chemical treatment. *Journal of Animal Science,* **61,** 363 (abstract)

Vintila, M., Damian, C., Burlacu, C., Gheorghiu, V. *et al.* (1980) Use of faeces in fermentation and in increasing the nutritive value of roughage. *Lucrarile Stiintifice ale Institutului de Cerelari Pentou Nutritive Animala,* **8,** 27–35 (Cited by *Nutrition Abstract and Review,* **50,** 6360)

Walker, H. G. (1984) Physical treatment. In *Straw and Other Fibrous By-products as Feed,* (ed. F. Sundstol and E. Owen). Developments in Animal and Veterinary Sciences, 14, Elsevier, pp. 79–105

Wanapat, M. (1986) Development of straw utilization as ruminant feed in Thailand. In *Rice Straw and Related Feeds in Ruminant Ration*, (ed. M. N. M. Ibrahim and J. B. Schiere). Department of Tropical Animal Production, Agricultural University, Wageningen, The Netherlands, pp. 86–98

Welsh Office Agricultural Department (WOAD) (1988) *Agricultural Census for Wales*, WOAD, Aberystwyth

Williams, J. E., McLaren, G. A., Smith, T. R. and Fahey, G. C. (1979) Soluble rumen liquor hemicellulose: composition and influence on *in vitro* rumen microbial protein synthesis. *Journal of Animal Science,* **49,** 163–168

Wilkinson, J. M. (1984) Ensiling with sodium hydroxide. In *Straw and Other Fibrous By-products as Feed,* (ed. F. Sundstol and E. Owen). Developments in Animal and Veterinary Sciences, 14, Elsevier, pp. 181–195

Woodruff, S. and Webber, L. R. (1933) A photomicrographic study of gelatinized wheat starch. *Journal of Agricultural Research,* **46,** 1099–1108

Woolford, M. K. (1984) *The Silage Fermentation*, Marcel Dekker, New York

Zadrazil, Fl. (1984) Microbial conversion of lignocellulose into feed. In *Straw and Other Fibrous By-products as Feed*, (ed. F. Sundstol and E. Owen). Developments in Animal and Veterinary Sciences, 14, Elsevier, pp. 276–292

Zimmer, E. and Wilkins, R. J. (eds) (1984) Efficiency of silage systems: a comparison between unwilted and wilted silages. *Landbauforschung Volkenrode,* **69,** 88 pp

Chapter 7

Recent developments in the nutrition of housed cattle

P. C. Thomas

In the field of animal nutrition practical developments take place as a consequence of advances in knowledge and understanding, improvements in manufacturing and processing technology, changes in economic and market conditions, and shifts in technological 'fashion'. However, whether these practical developments should be described as the introduction of 'new techniques' is questionable. 'New' implies some discontinuity in advances and developments – that there is some discrete point in time when a new technique is suddenly discovered or created. In animal nutrition this is rarely the case since research advances and practical developments tend to take place hand in hand, reflecting a very effective technology transfer from laboratory research to industrial practice. Since much large animal research is long term, elements of a new technique are generally in use in the industry before the technique has been conceived in its entirety. Thus new techniques are foreshadowed in a way that makes it difficult to pinpoint their precise moment of conception, and progress is better judged by reference to developments in scientific understanding and to changes in current concepts than to a list of specific practical procedures.

The nutrition of housed cattle represents a huge field of study. Therefore, within the limits of this chapter, comments will be confined to the major dietary nutrients conventionally described in terms of protein and energy and to dairy cows, which arguably are the most important class of housed cattle in the UK. The topics considered are: advances in protein nutrition; advances in energy nutrition; ration design and technology; feed allocation; and developments in nutritional concepts.

Advances in protein nutrition

So far as the protein nutrition of cattle is concerned the advances that have taken place over the past two decades can be summarized by reference to a few key publications. The Agricultural Research Council (ARC, 1965) highlighted the limitations of the traditional Digestible Crude Protein (DCP) system for rationing and proposed that the system should be replaced. However, the Available Protein (AP) system that was proposed as an alternative was not widely favoured and was not adopted for use in practice. It was in fact almost 10 years later that Miller (1973) made the first formal proposal of a new rationing system that would find support from practising nutritionists. Miller's ideas were subsequently developed and formed the basis of the recommendations for a new protein rationing system which was put forward by the Agricultural and Food Research Council (AFRC, 1980, 1984).

The new approach to protein rationing had its origins in work designed to provide a better understanding of the processes of protein digestion in the ruminant. These studies involved the use of animals cannulated in the abomasum or duodenum and terminal ileum (Faichney, 1975; MacRae, 1975). They provided quantitative information on the digestion of dietary protein in the rumen, on the ruminal synthesis of microbial protein and on the uptake of amino acids in the small intestine, and they provided a basis on which the digestion processes could be related quantitatively to the amount and composition of the diet.

The proposals put forward by Miller and incorporated into all later related systems of protein rationing involved the recognition:

(1) that there is an extensive breakdown of dietary protein in the rumen, and that this varies with the dietary protein source
(2) that there is a substantial synthesis of microbial protein in the rumen which depends on the dietary-supply of fermentable energy
(3) that the adequacy of the dietary protein supply to the animal should be assessed in terms of (a) the supply of nitrogen (N) to the rumen microorganisms and (b) the supply of amino acids to the host animal's small intestine
(4) that the deposition of protein in the animal's tissues, the secretion of protein in milk, etc. are functions of the intestinal amino acid supply and the efficiency of utilization of those amino acids.

An excellent summary of the quantitative information available on these processes and of the conceptual considerations underlying the development of new systems of protein rationing is provided by the National Research Council (1985).

In most countries of the world protein rationing systems based on the principles described in AFRC (1980, 1984) are now in common use, although in many countries, including the UK, the systems have yet to be officially adopted by the governmental agencies.

Since the early 1980s the main developments in protein rationing have reflected standardization and improvement in the intraruminal incubation techniques that are used as a basis for determining the rumen degradability of dietary proteins, refinements in technique to allow for variations between dietary protein sources in the intestinal digestibility of rumen-undegraded protein fractions and improvements in the mathematical description of the digestive processes in the rumen and intestines (Alderman and Jarrige, 1987). Alongside this there have been associated improvements in the accuracy of least cost ration formulation technique used by the feed industry and changes in ration design, to accommodate the new concepts of protein nutrition. In practice this has led to an increased use of high protein supplements and of protein sources of low rumen degradability.

Nonetheless, there is a number of areas in which the limitations on scientific understanding of protein digestion and metabolism remain a significant barrier to the further development of the new systems of protein rationing. Most notably there is still insufficient understanding of the regulation of microbial protein synthesis in the rumen, of the quantitative importance of endogenous secretions of N into the digestive tract and of the factors influencing the efficiency of utilization of absorbed amino acids. At the level of application of the systems there is also a continuing need for simple, rapid laboratory methods to determine the rumen

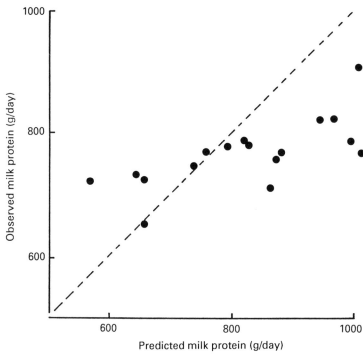

Figure 7.1 Observed milk protein outputs and milk protein outputs predicted on the basis of the AFRC (1984) protein rationing system. Figure is adapted from TCORN (1989) and is based on the results of Oldham *et al.* (1985) and Thomas *et al.* (1985) for cows given silage diets with supplements containing varying amounts of fishmeal and protein contents of 120–200 g/kg DM. The broken line represents the line of equivalence

degradable protein (RDP) and undegraded dietary protein (UDP) contents of feedstuffs (Alderman and Jarrige, 1987).

Of equal concern is the fact that the new systems have yet to be rigorously tested in practical feeding experiments with dairy cattle. Thus the accuracy and practical benefits of the new systems of rationing remain largely unquantified. Comparisons undertaken by AFRC (1980) showed that the new system had advantages over the DCP system, especially with diets containing highly degradable sources of protein. However, more recent studies have indicated that there are substantial errors and biases between measured milk protein outputs and those predicted by AFRC (1984) (Figure 7.1).

Advances in energy nutrition

Publication from ARC (1965) saw a major advance in systems for energy rationing in that it signalled the replacement of the Starch Equivalent system by the Metabolizable Energy (ME) system. The new system was not adopted in practice until 1975, and then in a somewhat modified form (Ministry of Agriculture Fisheries and Food (MAFF), 1975). However, its introduction was welcomed by industry, and the system rapidly acquired universal support as the basis for practical rationing.

AFRC (1980) proposed some modifications to the earlier ARC (1965) system but these were related to matters of detail rather than concept, and the principles put forward in 1980 were essentially those established in the earlier report. This could be taken to imply that few practical changes had taken place in the interim, but nothing could be further from the truth. Major developments had in fact taken place in the quantitative description of the ME content of common feeds through work undertaken at the Feed Evaluation Unit at the Rowett Research Institute, Aberdeen and the corresponding Agricultural Development and Advisory Service Unit, Drayton. An up-to-date compendium of this information has been published (MAFF, 1986). Additionally, there have been important recent developments in the prediction from laboratory analysis of the ME content of forages and compound concentrate feeds.

In studies reported by Kridis *et al.* (1987) and Barber, Offer and Givens (1989), it has been shown that currently used methods for the prediction of forage digestible organic matter (DOMD) or ME from laboratory determinations of modified acid detergent (MAD) fibre or lignin content are subject to large errors. However, a multifunctional prediction equation based on analysis of the feed's near infra-red (NIR) spectrum has been developed which gives rapid, accurate prediction of DOMD and thus ME content (Figure 7.2).

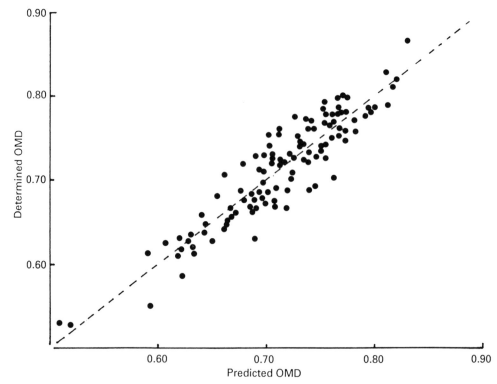

Figure 7.2 The relationship between determined *in vitro* values of organic matter digestibility (OMD) of silages and corresponding values predicted by near infra-red spectrometry. Results are from Barber *et al.* (1989). The broken line is the line of equivalence. The R^2 value for the relationship was 0.85

The prediction of the ME of commercial compounded concentrate feeds has been a matter of long-standing concern both in respect to quality control by the feed industry and in relation to customer assurance. Calorimetric studies with sheep fed at a maintenance level of feeding were undertaken by Wainman, Dewey and Boyne (1981) to establish an initial database from which predictive equations could be developed. These studies involved a total of 24 compounded feeds which were selected to cover a range of compositional specification including: two ranges of ether extract (20–39 g/kg DM and 50–70 g/kg DM); two ranges of crude fibre (40–60 g/kg DM and 80–120 g/kg DM); and three ranges of crude protein (120–149 g/kg DM, 150–179 g/kg DM and 180–209 g/kg DM). From an analysis of the results, UKASTA/ADAS/COSAC (1985) developed three recommended prediction equations (Table 7.1). These were adopted for practical use, but there was continued uncertainty about their application to diets containing high levels of added fat and/or added fibre. (As is discussed later these types of diets have now become commonplace.) Additionally, there was a frequently voiced concern about the accuracy of prediction equations developed from studies with sheep at a maintenance level of feeding when applied to lactating dairy cows fed at much higher feeding levels.

Table 7.1 Recommended equations for the prediction of metabolizable energy (ME) content from the chemical analysis of compounded feeds

Equation designation	Recommended use	Equation*	Standard deviation (MJ/kg DM)†
U1	Legal and voluntary	$ME = 11.78 + 0.0654\,CP + 0.0665\,EE^2 - 0.0414\,EE \times CF - 0.118\,TA$	0.36
U2	Reference purposes	$ME = 11.56 - 2.37\,EE + 0.030\,EE^2 + 0.030\,EE \times NCD - 0.034\,TA$	0.32
U3	Voluntary declaration	$ME = 13.83 - 0.488\,EE + 0.0394\,EE^2 \times CP - 0.0085\,MADF \times CP - 0.138\,TA$	0.35

* ME is MJ/kg DM, all other units are g/100 g DM. CP = crude protein; EE = ether extract; CF = crude fibre; TA = total ash; NCD = cellulase digestible organic matter in the dry matter following neutral detergent extraction; MADF = modified acid detergent fibre.

† Residual standard deviation taking account of between laboratory variances in chemical analysis of feed composition (*see* UKASTA/ADAS/COSAC, 1985).

To address these issues, a major collaborative study was undertaken (Thomas *et al.*, 1988). This involved calorimetric experiments with sheep at the maintenance level of feeding, which were conducted at the Rowett Research Institute, Aberdeen, and a linked programme of calorimetric work with dairy cows at production levels of feeding, which was conducted at the Hannah Research Institute, Ayr. In both studies the same forage and concentrate feeds were used. In the sheep experiments a total of 100 diets was studied. These included diets containing varying levels of added fats (0–90 g/kg) from each of five fat sources, and varying levels of added fibre (0–400 g/kg) from each of three fibre sources. The fat sources were palm acid oil, maize/soya oil, fat premix (FP1), Megalac and fat prills; the fat premix, Megalac and fat prills were all commercially prepared products selected on the basis of their physical, handling properties or because they were formulated to be 'rumen-protected', e.g. FP1. The fibre sources were straw, sodium hydroxide treated, nutritionally improved straw (NIS) and a 'digestible fibre' mixture of sugar beet pulp and citrus pulp.

From the 'matrix' of 100 diets, nine diets representative of the extremes that would normally be used for feeding dairy cows in practice were selected for examination in the cow experiments. These diets were based on two fibre sources, straw and sugar beet pulp/citrus pulp, and on two fat sources, palm acid oil and FP1, together with corresponding low-fibre and low-fat control treatments.

On completion of the sheep studies the data were combined with those from the earlier work of Wainman, Dewey and Boyne (1981) and subjected to comprehensive statistical analyses designed to establish the 'least error' relationships between feed ME content and chemical composition. The range of chemical constituents examined in this analysis was wide and a series of 'low error' equations was developed (Thomas *et al.*, 1988). However, none offered a significant advantage over a simple, two term equation (E3) which described the ME content of the feed in terms of the oil and NCD (*see below*) content. The equation was:

$$ME = 0.25 \ Oil + 0.14 \ NCD$$

where ME is metabolizable energy (MJ/kg DM), oil is total extractable lipid (method B;g/100 g DM) and NCD is cellulase digestible organic matter after neutral detergent extraction (g/100 g DM).

This equation was also found to have a satisfactory predictive capability for the ME content of feeds for dairy cows and produced figures in accord with those calculated from contemporary database values of the ME content of ingredient 'straight' feeds (Table 7.2).

Equation E3 has now been adopted as an industry standard in the UK (MAFF, 1989) and is finding widespread application.

Table 7.2 The metabolizable energy (ME) content (MJ/kg DM) of various compounded feeds determined in lactating cows, estimated from chemical composition using equation E3 and calculated from database values

Type of feed		Determined ME	Estimated ME from equation E3	ME from database values
Fibre source	*Fat source*			
None	None	12.43	12.47	12.43
None	Palm acid oil	14.42	14.24	13.73
None	FP1	13.49	13.65	13.81
Straw	None	10.35	10.10	9.94
Straw	Palm acid oil	11.87	11.65	11.39
Straw	FP1	11.55	11.34	11.48
Pulp*	None	12.43	12.39	12.01
Pulp*	Palm acid oil	13.30	14.16	13.49
Pulp*	FP1	13.40	13.99	13.79
Mean (with s.e.m.)		12.58 ± 0.40	12.67 ± 0.50	12.45 ± 0.46

From Thomas *et al.*, 1988.

* Sugar beet and citrus pulp.

Ration design and technology

In recent years there have been major changes in the design of rations for dairy cattle and in some related areas of feed technology. Underlying these changes has been a strong commercial pressure to increase the cost-effectiveness of milk production. In the UK, milk sales are limited by quota regulations on farm production and milk price is determined on the basis of differential payments for milk fat, protein and lactose. Thus, attention has been focused on the need to maximize the intake and efficiency of utilization of low-cost, home-grown forages and to improve milk compositional quality, the content of high-value fat and protein constituents in particular.

Silage technology

Grass silage is the main forage source for housed dairy cattle and not surprisingly the conservation process has been a major target for technological advance. Traditional silage additives, like molasses and chemical additives containing formic acid, sulphuric acid and formaldehyde alone or in combination have been in use for many years (Thomas and Thomas, 1985). However, recently there has been a major development in the use of biological additives, both bacterial inoculants and enzymes (celluloses and xylanases). A wide range of bacterial/enzyme additives is now available commercially. However, while most of these products have been tested in respect to their effects on silage fermentation quality, relatively few have been tested in animal performance trials, which provide the ultimate product evaluation.

At present, research results with silage inoculants do not allow firm conclusions to be drawn. Even where the same type of inoculant is used, animal responses to silage treatment appear to vary with circumstances. Thus on the basis of comparisons between inoculant silage and silage made either without additive or with formic acid additive, Gordon (1987) reported that silage intake and milk production were increased substantially through the use of *Lactobacillus plantarum*

Table 7.3 The responses of dairy cows in silage intake and milk yield to the use of *Lactobacillus plantarum* as a silage inoculant

Experimental comparison	Response in silage DM intake (%)		Response in milk yield (%)	
	(a)	(b)	(a)	(b)
1*	+12.5	+10.3	+9.5	+10.5
2	+ 0.8	− 6.1	−2.1	+ 0.8
3	+ 3.3	− 6.7	−1.5	− 1.9
4	+ 0.4	− 5.4	−0.1	− 0.7
5	− 0.5	− 6.3	+0.2	− 0.3
Mean	+ 3.3	− 3.9	+1.2	+ 1.68

* Results from Gordon (1987). Other results from Chamberlain, Thomas and Robertson (1987).

Responses are expressed as % change compared with a control silage (a) made without additive and (b) made with formic acid additive as silage inoculant (Table 7.3). However, benefits similar to *Lactobacillus plantarum* have not been observed in other, corresponding studies (Table 7.3). As yet there is no clear explanation for the divergence in these results.

Initial studies with enzyme silage additives containing cellulase and xylanase preparations indicated that the additives tended to reduce silage dry matter intake by cows but to maintain or increase milk yield, suggesting effects on the efficiency of feed utilization (Chamberlain, Thomas and Robertson, 1987). Recent studies with 'improved' enzyme mixtures have given more pronounced benefits including significant increases in silage intake and milk production (Table 7.4). Underlying

Table 7.4 The effect of enzyme silage additives on silage intake and milk production in dairy cows

	Control		Enzyme (E)		Enzyme (FS)		Enzyme (2FS)		s.e.m.
	L	H	L	H	L	H	L	H	
Feed intake									
Supplement (kgDM/day)	3.86	7.67	3.85	7.68	3.84	7.69	3.85	7.68	
Silage (kgDM/day)	9.57	8.32	9.64	8.04	10.59	9.06	10.33	8.84	0.15***
Total (kgDM/day)	13.43	15.99	13.49	15.72	14.43	16.75	14.18	16.52	0.15***
Milk production									
Milk yield (kg/day)	20.12	22.65	20.34	22.42	20.18	22.70	21.14	22.09	0.2**
Fat (g/kg)	38.6	39.0	37.8	37.2	38.3	38.4	36.9	35.7	0.4
Protein (g/kg)	29.7	32.1	29.6	32.1	30.2	31.5	30.5	32.1	0.1
Lactose (g/kg)	47.0	47.2	47.1	46.8	46.7	46.9	47.1	47.1	0.1
Silage digestibility									
DOMD (g/kg)†	609		588		624		635		

Results are from Chamberlain and Robertson (1988) and refer to a control silage made without additive and to three silages made from the same grass but with the addition of cellulose/hemicellulose mixtures. There were two enzyme mixtures (E and FS). Rates of application were: E 0.2 l/t; FS 0.3 l/t; 2FS 0.6 l/t. Each silage was given together with a high protein concentrate supplement given at either a low (L) or a high (H) level of feeding.

* s.e.m., standard error of the mean refers to comparisons between silages at corresponding levels of supplement intake: ** $P<0.01$; *** $P<0.001$.

† DOMD (g/kg), digestible organic matter in the silage dry matter measured at a maintenance level of feeding in sheep.

these production responses are the effects of the enzymes in increasing the digestibility of silage organic matter (Chamberlain and Robertson, 1988) and in enhancing the efficiency of microbial protein synthesis in the rumen (McAllan, personal communication, 1988). The developments in enzyme technology appear to offer the exciting prospect that silage crops might be cut at a relatively mature stage when yield is heavy, but digestibility and intake are relatively low, and then 'upgraded' by enzyme treatment during storage in the silo.

Composition of supplementary feeds

There has been a progressive recognition over time that the amount and composition of the supplementary feeds given to cows influences silage intake, milk yield and milk fat and protein content. However, it is only comparatively recently that this has begun to have a major impact on ration design. This has led to a more creative approach to ration formulation which rejects the classical principle that supplementary feeds can be equated simply on the basis of their ME and digestible crude protein contents and emphasizes the importance of 'complementary feeds'.

Replacement or modification of starch

At the levels of concentrate supplementation used in practice (3–12 kg/day), increased allowances of starchy, cereal based feeds lead to a reduction in silage intake and a depression in milk fat content; both effects becoming more pronounced as the level of concentrate allowance is raised. These effects have their origin in the influence of dietary starch on ruminal cellulolysis and on the ratio of acetic and butyric acids:propionic acid formed during rumen fermentation (Thomas and Rook, 1977). Cellulolysis and the acetic + butyric acid:propionic acid ratio are reduced less by supplements containing digestible fibre feeds (sugar beet pulp, citrus pulp etc.) than by those containing starch. In recent years this has led to an increased use of fibrous feeds in cattle concentrates. Experiments have shown that, compared with starchy, cereal concentrates, digestible fibre concentrates lead to a higher silage intake and to a higher milk fat content and/or milk yield (Castle, Gill and Watson, 1981; Sloan, Rowlinson and Armstrong, 1986).

The effects of cereal supplements on silage intake have also been modified by 'cross-linking' the cereal starch with acidic formaldehyde solution to slow the rate of starch fermentation in the rumen (Kassem et al., 1987). In experiments with cows given silage and barley diets, treatment of barley has been shown to increase silage intake, milk yield and milk protein content but reduce milk fat content (Table 7.5). The effects on milk yield and fat content may be explained in part by an increased passage of starch to the small intestine with the treated barley supplements, or by the fact that formaldehyde treatment also 'cross-links' cereal

Table 7.5 The effect of formaldehyde treatment of barley grain on silage intake and milk production of multiparous dairy cows given supplements with or without added fishmeal

	Untreated barley		Treated barley		SE diff
	− Fishmeal	+ Fishmeal	− Fishmeal	+ Fishmeal	
Feed intake					
Supplement (kgDM/day)	5.85	5.89	5.83	5.87	–
Silage (kgDM/day)	8.33	9.51	8.92	9.71	0.28
Total (kgDM/day)	14.18	15.40	14.75	15.58	0.28
Milk production					
Milk yield (kg/day)	19.40	21.05	21.12	22.40	0.45
Fat content (g/kg)	46.40	45.20	43.60	43.00	2.30
Protein content (g/kg)	30.70	31.00	31.20	31.80	0.90
Lactose content (g/kg)	48.50	48.40	48.20	48.00	0.36

From Kassem, *et al.*, 1987.

protein, increasing its passage to the post-ruminal gut (Van Ramshorst and Thomas, 1988).

Inclusion of sugars

The efficiency of microbial synthesis in the rumen of cattle given silage diets is low. This leads to a high absorption of ammonia from the rumen, to a reduced passage of microbial protein to the small intestine, and to potential constraints on animal production arising from limitations on amino acid supply (Thomas and Chamberlain, 1982). There are a number of clear indications that this situation could be improved through the dietary provision of readily fermentable sources of carbohydrate and that sugars may have advantages over starchy supplements (Chamberlain et al., 1985).

This may in part reflect differences in the rates of ruminal fermentation of sugars and starch, but it should also be noted that starch supplements increase the numbers of protozoa in the rumen, and this has an adverse effect on the efficiency of ruminal nitrogen utilization (Chamberlain et al., 1985).

Supplementation of silage diets with sugars has been shown substantially to increase microbial protein synthesis in the rumen and duodenal protein flow (Huhtanen, 1987; Rooke, Lee and Armstrong, 1987). Moreover, the effects are enhanced when the sugar is given in combination with a protein source or a protein source and sodium bicarbonate (Rooke, Lee and Armstrong, 1987; Newbold, Thomas and Chamberlain, 1988; Newbold, Chamberlain and Thomas, 1989). An example of the synergistic effects of the supplements is shown in Table 7.6. As can be seen the major influence of the sodium bicarbonate is to increase the rate of microbial protein synthesis in the rumen.

Table 7.6 The effects of dietary supplements of a mixture of soya bean meal/fishmeal (SF) or of sodium bicarbonate given alone or in combination on duodenal nitrogen flow in sheep given diets of silage and molasses

	Dietary supplement				
	None	SF	NaHCO₃	SF + NaHCO₃	SE diff
Nitrogen intake (g/day)	22.6	29.9	22.6	29.9	
Duodenal total nitrogen (g/day)	16.8	20.0	20.0	25.0	2.2
Duodenal microbial nitrogen (g/day)	9.2	11.0	15.7	20.4	3.3

Results are from Newbold, Chamberlain and Thomas, 1989.

These observations have led to the concept that a conventional allowance of cereal-based supplement might be replaced by a lower allowance of sugar-based supplement designed specifically to complement grass silage. To examine this idea, Rae, Thomas and Reeve (personal communication, 1988) compared 4.3 kg/day of a conventional cereal-based concentrate with 3.6 kg/day of a concentrate containing molasses, soya bean meal, fish meal and sodium carbonate (45: 33: 11: 11) each given to cows receiving silage *ad libitum*. Under the conditions of the experiment animals receiving the 'sugar' supplement consumed 1.3 kg DM/day more silage and produced 0.9 kg/day more milk, with a reduction in milk fat content (38.2 *versus*

41.7 g/kg) and an increase in milk protein content (32.4 *versus* 31.0 g/kg). Thus the approach appears to have potential, although it should be noted that responses to sugar supplements will by their nature depend on the nitrogen content of the silage in the diet.

Protein and amino acid inclusions

There is no doubt that the publication of the AFRC (1980, 1984) protein rationing system served to focus attention on that aspect of cattle nutrition. However, changes in approach to protein feeding in practice were beginning to take place well before the new system was in widespread use. As early as the mid 1970s there was evidence that under specified circumstances cows would respond in milk production to dietary protein allowances well above those recommended by contemporary nutritional standards. As understanding of these responses grew it became apparent that they involved a combination of effects on forage intake, on digestibility of organic matter in the rumen and on amino acid supply to the small intestine (Oldham, 1984).

It is now common practice for dairy cows to be given allowances of protein above recommended standards, and particularly for rations to be formulated with an 'excess' of rumen-undegradable protein. This approach is based mainly on the results of feeding experiments which have demonstrated increased milk production in response to dietary inclusions of animal protein sources, such as fishmeal and blood meal (e.g. *see* Table 7.7). However, it should be noted that the economics of rationing to protein 'excess' require further research, as does the definition of the optimum level of protein allowance.

Table 7.7 The effect of dietary supplements of rumen-protected methionine and lysine or of a low rumen-degradable protein source on feed intake and milk production in dairy cows

	Dietary supplement				
	None	*Methionine + lysine†*	*Methionine + lysine‡*	*Protein source§*	*SE diff*
Feed intake (kgDM/day)*	17.01	16.71	16.32	17.17	0.4
Milk yield (kg/day)	19.84	19.02	19.41	23.04	0.66
Fat (g/kg)	43.0	46.3	48.6	41.2	1.7
Protein (g/kg)	32.9	33.5	34.0	32.9	0.4
Lactose (g/kg)	46.0	45.6	47.0	47.0	0.5

Results are from Girdler, Thomas and Chamberlain, 1987.

* Cows received *ad libitum* a complete mixed diet of silage and barley together with a concentrate allowance of 1.1 kg/day which provided the dietary supplement in a mixture with barley.
† Methionine 12 g/day, lysine 18 g/day.
‡ Methionine 12 g/day, lysine 36 g/day.
§ Protein source with a mixture of fishmeal, blood meal and meat and bone meal.

From an analysis of the amino acids entering the small intestine Thomas and Chamberlain (1982) concluded that methionine and lysine were the first limiting amino acids for milk production in cows given silage–barley diets. Subsequently, a number of studies was undertaken to investigate the use of dietary supplements of specific amino acids provided in a form 'protected' from degradation in the rumen

(Shamoon, 1983; Girdler, Thomas and Chamberlain, 1987). In these studies methionine or methionine and lysine resulted in small responses in milk fat and/or milk protein content, but in no instance was it possible to elicit milk production responses similar to those observed with low-degradability, high methionine and lysine protein sources (Table 7.7). Corresponding results with protected methionine and lysine have also been reported by Sloan and Thomas (unpublished data, 1989) and with methionine-hydroxy analogue (Gordon and Unsworth, 1986). On the basis of these results the development of specific dietary amino acid supplements for cattle can only be regarded as a matter for further research.

Fats and protected fats

A major recent development in the formulation of cattle feeds has been the increasing use of fats and oils. It is now not uncommon for dairy concentrate feeds to contain 60–80 g/kg of added fat, which is introduced partly through in-mixing and partly through spray treatment of mixed feed. There has also been a substantial increase in the use of 'dry' fat products. These are prepared from 'non-polar' saturated triglycerides and/or 'polar' components such as fatty acids, calcium salts of fatty acids or proteins. During their preparation the products are stabilized as powders or prills through spray cooling or spray drying technologies which result in a particle with a 'non-polar' core surrounded by a 'polar' coating. Due to their structure many of these products have a degree of 'ruminal protection'. This is important not so much from the stand point of avoiding the biohydrogenation of fatty acids in the rumen (Fogerty and Johnson, 1980), but more because it reduces the adverse effects of the fats on rumen microbial activity, including the effects on cellulolysis and rumen volatile fatty acid production (Storry and Brumby, 1980).

The use of fat supplements in diets for dairy cows has been extensively reviewed (Storry, 1980; Banks, Clapperton and Steele, 1983; Clapperton and Steele, 1983; Palmquist, 1984; Czerkawski and Clapperton, 1984; Clapperton and Banks, 1985). These reviews have shown that the effects of the fats on milk production vary with type of fat, form of inclusion in the diet, level of inclusion, and frequency of

Table 7.8 The effect of dietary inclusions of palm acid oil or of protected palm acid oil (FP1) on milk yield and composition in cows receiving concentrates containing no added fibre, straw or sugar beet pulp/citrus pulp together with grass silage *ad libitum*

Type of feed		Feed intake (kgDM/day)	Milk yield (kg/day)	Milk fat (g/kg)	Milk protein (g/kg)
Fibre source	Fat source				
None	None	16.61	19.9	40.1	33.7
None	Palm acid	15.53	21.0	42.3	30.9
None	FP1	15.50	24.4	35.2	29.3
Straw	None	16.78	19.3	41.9	32.3
Straw	Palm acid	16.14	20.9	41.3	30.6
Straw	FP1	15.78	21.7	37.0	29.4
Pulp	None	16.14	21.7	43.4	32.5
Pulp	Palm acid	16.18	21.1	40.0	31.2
Pulp	FP1	16.07	22.4	34.7	30.0
SE of diff		0.27	1.1	1.5	0.5

Results are from Thomas and Robertson, 1987.

feeding. They have also provided some guidelines on the circumstances and the ways in which fat supplementation techniques may most satisfactorily be adopted. Less well recognized is the fact that there are subtle interactions between added fats and the other components of the diet so that the precise response to the fat supplements in terms of increase in milk yield and/or change in milk fat and protein content is difficult to predict. In practice, this significantly detracts from the usefulness of fat supplementation as a technique. The problems are well illustrated by reference to an experiment which was conducted at the Hannah Research Institute as part of the study on the nutritional value of various fat and fibre sources in the diet of the dairy cow, (see p. 91). In this experiment cows were given grass silage *ad libitum* together with 8 kg/day of one of nine concentrate diets containing different levels and types of added sources of fat and added sources of fibre. Fat sources were included at a rate of 60 g/kg and fibre sources at 400 g/kg. The range of experimental treatments and the experimental results are summarized in Table 7.8. As can be seen, the response to the fat supplements varied not only with the type of fat that was used but also with the level and type of fibre included in the concentrate feed. Moreover, the effects of the protected fat supplement consistently and unexpectedly led to a reduction in milk fat content and an associated increase in milk yield.

Feed allocation

In the 1970s opinions on the nutritional management of dairy cows throughout lactation were based heavily on the 'classical' experiments conducted by Broster and his colleagues at the National Institute for Research in Dairying (Broster and Thomas, 1981; Broster and Broster, 1984).

These experiments, which were undertaken mainly with first lactation cows given hay-based diets, led to the following views:

(1) the cow's feed intake in early lactation is physiologically constrained within narrow boundaries, limiting the proportion of forage feeds that can be incorporated in the diet
(2) lactation performance is determined by 'peak' milk yield, the peak value signifying the expression of the cow's 'lactation potential'
(3) mobilization of body tissue, mainly adipose tissue, in early lactation is:
    (a) a function of the genetic make-up of the cow and
    (b) dependent on the level of energy intake pre- and post-calving.

However, as knowledge has been acquired over the past decade the above views have been progressively qualified and modified to a point where the narrow maxims that were previously propounded are no longer tenable. At both the experimental and practical level there is now irrefutable evidence of complex and interdependent relationships between physiological state of the cow, feed intake, milk yield and body tissue mobilization or deposition. Most important perhaps has been the recognition that the relationships are influenced by the *ad libitum* availability of high-quality forages, by the level and type of protein in the diet, by its interaction with the level of energy intake and by the carbohydrate composition of the concentrate feed (Ørskov, Grubb and Kay, 1977; Ørskov, Reid and MacDonald, 1981; Lees, Garnsworthy and Oldham, 1982). This has led to the development of 'flat-rate' systems of feeding (Leaver, 1988), to an emphasis on body condition score as an aid to feeding management (Garnsworthy, 1988) and to the use of rations of high digestible fibre and UDP content, designed to promote both forage intake and body tissue mobilization in early lactation.

Developments in nutritional concepts

The AFRC (1980, 1984) systems for energy and protein rationing adopt a similar approach in that they ascribe energy and protein values to feeds and assess the animal's needs through a factorial calculation of requirement for maintenance, milk production and tissue gain. The systems embody the assumption that the characteristics of the animal impose a predetermining influence on the animal's performance (e.g. on its milk yield and composition) and that the supply of energy and protein in the diet simply serves to allow that level of performance to be achieved. However, this assumption is inconsistent with research findings.

Numerous studies have shown that cows respond to an increased energy intake by increasing both milk yield and bodyweight gain. The increase in energy intake produces a negatively curvilinear response in milk yield, even at levels of energy supply well beyond those indicated by recommended feeding standards, and there is an associated positive curvilinear response in weight gain (Yates, Boyd and Petit, 1942; Blaxter, 1967). A major factor determining the quantitative nature of the milk yield response is the current milk yield, implying that responsiveness is a function of previous as well as current levels of nutrition (Burt, 1957; Broster and Broster, 1984). Furthermore, as has been identified earlier, changes in the chemical composition of the diet are important. They influence the partition of energy use between milk production and bodyweight gain and thus modulate the shape of the energy response curves (Broster, Sutton and Bines, 1979; Thomas and Chamberlain, 1984).

Alongside this there are similar considerations with regard to changes in dietary protein supply, responses being observed up to and beyond the level of recommended dietary standards, and including effects on milk yield and bodyweight gain (Oldham, 1984; Whitelaw et al., 1986) or in early lactation bodyweight loss (Ørskov, Grubb and Kay, 1977; Ørskov, Reid and MacDonald, 1981).

Similarly, there is ample evidence to show that milk composition is not solely an attribute of the cow. There are major influences of diet composition on milk fat and protein content which are regulated through dietary effects on the production of acetic acid, propionic acid, butyric acid in the rumen and glucose, amino acids and long-chain fatty acids in the small intestine (Thomas and Martin, 1988). The mixture of products of digestion may influence milk secretion and body tissue synthesis directly through precursor–product relationships depending on the substrate supply from the gut. The balance of substrates absorbed also plays an important role in determining the relative rates of secretion of key metabolic hormones such as insulin and somatotrophin, which themselves serve to regulate nutrient use for milk secretion and tissue synthesis (Hart, 1983).

The limitations of feeding systems based on requirements for energy and protein have increasingly been recognized and recently there has been a major conceptual swing in favour of the development of alternative systems. These are based on concepts of animal response to nutrient supply rather than on concepts of 'requirements', and nutrient supply is expressed in terms that reflect the production

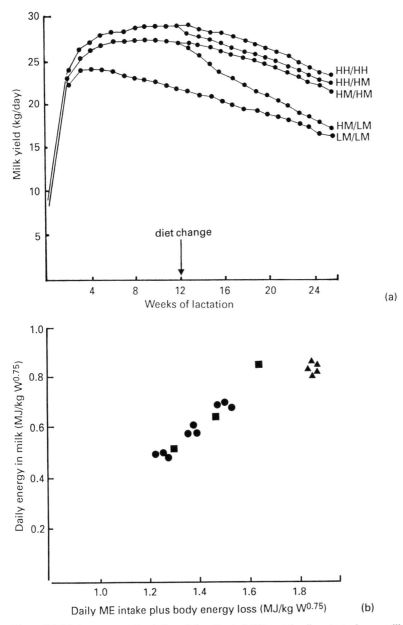

Figure 7.3 (*a*) A computer simulation of the effect of different feeding strategies on milk yield in the cow. Dietary treatments were either held constant for 26 weeks of lactation or were changed at week 12 of lactation. Dietary treatments were based on a forage:concentrate diet (50:50). The treatments were: HH, diet containing 180 g crude protein/kg and given at 8 kg/day plus 0.33 kg/kg milk yield; HM, diet containing 150 g crude protein/kg and given at 8 kg/day plus 0.33 kg/kg milk yield; LM, diet containing 150 g crude protein/kg and given at 5 kg/day plus 0.33 kg/kg milk yield. (*b*) Relationship between milk energy loss and daily intake of metabolizable energy (ME) plus energy mobilized from body tissues. ●, Values derived from computer simulation for cows in weeks 17–28 of lactation. ▲, Values derived from computer simulation for cows in weeks 4–9 of lactation. ■, Values determined experimentally in cows over the first 8 weeks of lactation. (After Baldwin *et al.*, 1987)

of products of digestion. Since the latter include both 'energy' constituents and amino acids, the traditional distinction between energy and protein feeding systems disappears, as both are incorporated into a single framework.

In a scientific sense the development of fully satisfactory new systems will depend on the growing understanding of quantitative aspects of digestion and metabolism in cattle. However, at the technical level the important breakthrough has been in computer modelling and simulation techniques. Research models designed to simulate the processes of digestion and metabolism in the cow have already been constructed (Baldwin *et al.*, 1987; Baldwin, France and Gill, 1987; Baldwin, Thornley and Beever, 1987) and have provided encouraging results (Figure 7.3). Less complex models designed for practical application are also being developed (Oldham, 1988; Webster, Dewhurst and Waters, 1988).

There is therefore a real prospect for the development of predictive feeding systems which relate animal performance to diet supply in quantitative terms. These systems necessarily will need to encompass or incorporate response predictions of feed intake, production of products of digestion and utilization of those products for the secretion of individual milk constituents and for body fat and protein synthesis (Oldham and Emmans, 1988). However, it has been concluded that given appropriate levels of research and development this objective is scientifically achievable in a reasonably short time (AFRC Technical Committee on Responses to Nutrients (TCORN), 1989).

The introduction of this new type of feeding system in practice would have very substantial benefits. It would promote the optimum use of feed resources, allow milk quality to be tailored to meet consumer requirements and milk constituent prices, and reduce the uncertainties associated with changes in feeding management which bedevil the dairy industry in practice. In short, it would replace a great deal of the 'nutritionist's art' with objective science!

Acknowledgements

I am grateful to a number of colleagues who have kindly allowed me to quote from work which is unpublished or published only in abstract form.

References

AFRC Technical Committee on Responses to Nutrients (1989) Responses in the yields of milk constituents to the intake of nutrients by dairy cows. *Nutrition Abstracts and Reviews* (in press)

Agricultural and Food Research Council (1980) *The Nutrient Requirements of Ruminant Livestock.* Commonwealth Agricultural Bureaux, Farnham Royal

Agricultural and Food Research Council (1984) *The Nutrient Requirements of Ruminant Livestock*, suppl. no. 1. Commonwealth Agricultural Bureaux, Farnham Royal

Agricultural Research Council (1965) *The Nutrient Requirements of Farm Livestock no 2.* Ruminants. ARC, London

Alderman, G. and Jarrige, R. (1987) *Protein Evaluation of Ruminant Feeds.* Committee of European Communities, Luxembourg

Baldwin, R. L., France, J., Beever, D. E., Gill, M. *et al.* (1987) Metabolism in the lactating cow. III Properties of mechanistic models suitable for evaluation of energetic relationships and factors involved in the partition of nutrients. *Journal of Dairy Research,* **54**, 133–145

Baldwin, R. L., France, J. and Gill, M. (1987) Metabolism of the lactating cow. 1. Animal elements of a mechanistic model. *Journal of Dairy Research,* **54**, 77–105

Baldwin, R. L., Thornley, J. H. M. and Beever, D. E. (1987) Metabolism in the lactating cow. II Digestive elements of a mechanistic model. *Journal of Dairy Research,* **54,** 107–131

Banks, W., Clapperton, J. L. and Steele, W. (1983) Feeding fat to dairy cows. *Hannah Research Institute Report – 1982,* pp. 75–83

Barber, G. B., Givens, D. I., Kridis, M. S., Offer, N. W. *et al.* (1989) Prediction of organic matter digestibility of grass silage. *Animal Feed Science and Technology* (in press)

Barber, G. D., Offer, N. W. and Givens, D. I. (1989) Predicting the nutritive value of silage. In *Recent Advances in Animal Nutrition – 1989,* (ed. W. Haresign and D. J. A. Cole). (in press)

Blaxter, K. L. (1967) The feeding of dairy cows for optimal production. *George Scott Robertson Memorial Lecture,* Belfast, Queens University

Broster, W. H. and Broster, V. J. (1984) Long-term effects of plane of nutrition on the performance of the dairy cow. *Journal of Dairy Research,* **51,** 149–196

Broster, W. H., Sutton, J. D. and Bines, J. A. (1979) Concentrate forage ratios for high-yielding dairy cows. In *Recent Advances in Animal Nutrition – 1978,* (ed. W. Haresign and D. Lewis). Butterworths, London, pp. 99–126

Broster, W. H. and Thomas, C. (1981) Influence of level and pattern of concentrate input on milk output. In *Recent Advances in Animal Nutrition – 1981,* (ed. W. Haresign). Butterworths, London, pp. 49–69

Burt, A. W. A. (1957) The influence of level of feeding during lactation upon the yield and composition of milk. *Dairy Science Abstracts,* **19,** 435–454

Castle, M. E., Gill, M. S. and Watson, J. N. (1981) Silage and milk production: a comparison between barley and dried sugar beet pulp as silage supplements. *Grass and Forage Science,* **36,** 319–324

Chamberlain, D. G. and Robertson, S. (1988) The effects of various enzyme mixtures and silage additives on feed intake and milk production in dairy cows. *British Grassland Society Occasional Symposium* (in press)

Chamberlain, D. G., Thomas, P. C. and Robertson, S. (1987) The effect of formic acid, bacterial inoculant and enzyme additives on feed intake and milk production in cows given silage of high or moderate digestibility with two levels of supplementary concentrates. In *Proceedings of the Eighth Silage Conference,* (ed. C. Thomas). AFRC Institute of Grassland and Animal Production, Hurley, pp. 31–32

Chamberlain, D. G., Thomas, P. C., Wilson, W., Newbold, C. J. *et al.* (1985) The effects of carbohydrate supplements on ruminal concentrations of ammonia in animals given diets of grass silage. *Journal of Agricultural Sciences, Cambridge,* **104,** 331–340

Clapperton, J. L. and Banks, W. (1985) Factors affecting the yield of milk and its constituents particularly fatty acids when dairy cows consume diets containing added fat. *Journal of the Science of Food and Agriculture,* **36,** 1205–1211

Clapperton, J. L. and Steele, W. (1983) Fat supplementation in animal production – ruminants. *Proceedings of the Nutrition Society,* **42,** 343–350

Czerkawski, J. W. and Clapperton, J. L. (1984) Fats as energy yielding compounds in the ruminant diet. In *Fats in Nutrition,* (ed. J. Wiseman). Butterworths, London. pp. 249–263

Faichney, G. J. (1975) The use of markers to partition digestion within the gastrointestinal tract of ruminants. In *Digestion and Metabolism in the Ruminant,* (ed. I. W. McDonald and A. C. I. Warner). University of New England Publishing Unit, Armidale. pp. 227–291

Fogerty, A. C. and Johnson, A. R. (1980) Influence of nutritional factors on the yield and content of milk fat: protected polyunsaturated fat in the diet. In *Factors Affecting the Yields and Contents of Milk Constituents of Commercial Importance.* International Dairy Federation, Brussels. pp. 96–104

Garnsworthy, P. C. (1988) The effect of energy reserves at calving on performance of dairy cows. In *Nutrition and Lactation in the Dairy Cow,* (ed. P. C. Garnsworthy). Butterworths, London. pp. 157–170

Girdler, C. P., Thomas, P. C. and Chamberlain, D. G. (1987) Responses to dietary rumen-protected amino acids and abomasally infused amino acids in lactating cows given silage diets. In *Proceedings of the Eighth Silage Conference,* (ed. C. Thomas). AFRC Institute of Grassland and Animal Production, Hurley. pp. 73–74

Gordon, F. J. (1987) An evaluation of an inoculant as an additive for grass silage being offered to dairy cattle. In *Proceedings of the Eighth Silage Conference,* (ed. C. Thomas). AFRC Institute of Grassland and Animal Production, Hurley. pp. 19–20

Gordon, F. J. and Unsworth, E. F. (1986) The effects of silage harvesting system and supplementation of silage based diets by protein and methionine hydroxy analogue on the performance of lactating cows. *Grass and Forage Science,* **41**, 1–8

Hart, I. C. (1983) Endocrine control of nutrient partition in lactating ruminants. *Proceedings of the Nutrition Society,* **42**, 181–194

Huhtanen, P. (1987) The effect of carbohydrate supplements on the utilization of grass silage diets. Department of Animal Production, University of Helsinki, Helsinki

Kassem, M. M., Thomas, P. C., Chamberlain, D. G. and Robertson, S. (1987) Silage intake and milk production in cows given barley supplements of reduced ruminal degradability. *Grass and Forage Science,* **42**, 175–183

Kridis, M. S., Barber, D. G., Offer, N. W., Murray, I. *et al.* (1987) Prediction of organic matter digestibility of grass silages: recent developments. In *Proceedings of the Eighth Silage Conference,* (ed. C. Thomas). AFRC Institute of Grassland and Animal Production, Hurley. pp. 53–54

Leaver, J. D. (1988) Level and pattern of concentrate allocation to dairy cows. In *Nutrition and Lactation in the Dairy Cow,* (ed. P. C. Garnsworthy). Butterworths, London. pp. 315–326

Lees, J. A., Garnsworthy, P. C. and Oldham, J. D. (1982) The response of dairy cows in early lactation to supplements of protein given with rations designed to promote different patterns of rumen fermentation. In *Forage Protein in Ruminant Animal Production,* (ed. D. J. Thompson, D. E. Beever and R. G. Gunn). *Occasional Publication of the British Society of Animal Production, no 6.* BSAP, Thames Ditton, pp. 157–159

Macrae, J. C. (1975) The use of re-entrant cannulae to partition digestive function within the gastrointestinal tract of ruminants. In *Digestion and Metabolism in the Ruminant,* (ed. I. W. MacDonald and A. C. I. Warner). University of New England Publishing Unit, Armidale. pp. 261–276

MAFF (1975) *Energy Allowances and Feeding Systems for Ruminants,* Bulletin 33. Her Majesty's Stationery Office, London. p. 85

MAFF (1986) Feed composition. *Ministry of Agriculture, Fisheries and Food Standing Committee on Food Composition.* Chalcombe Publications, Marlow

MAFF (1989) *The Prediction of the Metabolizable Energy Values of Compound Feeding Stuffs for Ruminant Animals.* Ministry of Agriculture Fisheries and Food, London

Miller, E. L. (1973) Evaluation of foods as sources of nitrogen and amino acids. *Proceedings of the Nutrition Society,* **32**, 79–84

National Research Council (1985) *Ruminant Nitrogen Usage.* National Academy Press, Washington. p. 138

Newbold, J. C., Chamberlain, D. G. and Thomas, P. C. (1989) Effect of dietary supplements of sodium bicarbonate with or without additional protein on the utilization of nitrogen in the rumen of sheep receiving a silage-based diet. *Journal of Agricultural Science, Cambridge,* (in press)

Newbold, J. C., Thomas, P. C. and Chamberlain, D. G. (1988) Effect of dietary supplements of sodium bicarbonate on the utilization of nitrogen in the rumen of sheep receiving a silage-based diet. *Journal of Agricultural Science, Cambridge,* **110**, 383–386

Oldham, J. D. (1984) Protein-energy relationships in dairy cows. *Journal of Dairy Science,* **67**, 1090–1114

Oldham, J. D. (1988) Nutrient allowances for ruminants. In *Recent Advances in Animal Nutrition – 1988,* (ed. W. Haresign and D. J. M. Cole). Butterworths, London. pp. 147–166

Oldham, J. D. and Emmans, G. E. (1988) Predicting responses to protein and energy yielding nutrients. In *Nutrition and Lactation in the Dairy Cow* (ed. P. C. Garnsworthy), Butterworths, London, pp. 76–96

Oldham, J. D., Phipps, R. H., Fulford, R. J., Napper, D. J. *et al.* (1985) Responses of dairy cows to rations varying in fish meal or soya bean meal content in early lactation. *Animal Production,* **40**, 519

Ørskov, E. R., Grubb, D. A. and Kay, R. N. B. (1977) Effect of postruminal glucose or protein supplementation on milk yield and composition in Friesian cows in early lactation. *Journal of Nutrition,* **38**, 397–405

Ørskov, E. R., Reid, G. W. and Macdonald, I. (1981) The effects of protein degradability and food intake on milk yield and composition in cows in early lactation. *British Journal of Nutrition,* **45**, 547–555

Palmquist, D. (1984) Use of fats in diets for lactating cows in early lactation. In *Fats in Animal Nutrition*, (ed. J. Wiseman). Butterworths, London. pp. 357–381

Rooke, J. A., Lee, W. H. and Armstrong, D. G. (1987) The effects of intraruminal infusions of urea, casein and glucose syrup and a mixture of casein and glucose syrup on nitrogen digestion in the rumen of cattle receiving grass silage diets. *British Journal of Nutrition*, **57**, 89–98

Shamoon, S. A. (1983). Amino acid supplements for ruminant farm livestock with special reference to methionine. *PhD Thesis*, University of Glasgow

Sloan, B. K., Rowlinson, P. and Armstrong, D. G. (1986) The influence of concentrate energy source on dairy cows performance. *Animal Production*, **42**, 434 (abstract)

Storry, J. E. (1980) Influence of nutritional factors on the yield and content of milk fat: non-protected fat in the diet. In *Factors Affecting the Yields and Contents of Milk Constituents of Commercial Importance*. International Dairy Federation, Brussels. pp. 105–125

Storry, J. E. and Brumby, P. E. (1980) Influence of nutritional factors on the yield and content of milk fat: protected non-polyunsaturated fats in the diet. In *Factors Affecting the Yields and Contents of Milk Constituents of Commerical Importance*. International Dairy Federation, Brussels. pp. 105–125

TCORN, (1989) Responses in the yields of milk constituents to the intake of nutrients by dairy cows. *Report of the Agricultural and Food Research Council Technical Committee on Response to Nutrients, Nutrition Abstracts and Reviews*, (in press)

Thomas, C., Aston, K., Daley, S. R. and Beever, D. E. (1985) The effect of level and pattern of protein supplementation on milk output. *Animal Production*, **40**, 519

Thomas, C. and Thomas, P. C. (1985) Factors affecting the nutritive value of grass silages. In *Recent Advances in Animal Nutrition – 1985*, (ed. W. Haresign and D. J. A. Cole). Butterworths, London. pp. 223–257

Thomas, P. C. and Chamberlain, D. G. (1982) Utilization of silage nitrogen. In *Forage Protein Conservation and Utilization*, (ed. T. W. Griffiths and M. F. Maguire). Commission of the Economic Communities, Brussels. pp. 121–146

Thomas, P. C. and Chamberlain, D. G. (1984) Manipulation of milk composition to meet market needs. In *Recent Advances in Animal Nutrition – 1984*, (ed. W. Haresign and D. J. Cole). Butterworths, London. pp. 219–243

Thomas, P. C. and Martin, D. A. (1988) The influence of nutrient balance on milk yield and composition. In *Nutrition and Lactation in the Dairy Cow*, (ed. P. C. Garnsworthy). Butterworths, London. pp. 97–118

Thomas, P. C. and Robertson, S. (1987) The effect of lipid and fibre source and content on silage intake, milk production and energy utilization. In *Proceedings of the Eighth Silage Conference*, (ed. C. Thomas). Institute of Grassland and Animal Production, Hurley, pp. 173–174

Thomas, P. C., Robertson, S., Chamberlain, D. G., Livingstone, R. M. *et al.* (1988) Predicting the metabolizable energy (ME) content of compound feed for ruminants. In *Recent Advances in Animal Nutrition – 1988*, (ed. W. Haresign and D. J. A. Cole). Butterworths, London. pp. 127–146

Thomas, P. C. and Rooke, J. A. F. (1977) Manipulation of rumen fermentation. In *Recent Advances in Animal Nutrition – 1977*, (ed. W. Haresign and D. Lewis). Butterworths, London. pp. 83–109

UKASTA/ADAS/COSAC (1985) Prediction of the energy value of compound feeds. *Report of the UKASTA/ADAS/COSAC Working Party*, United Kingdom Agricultural Suppliers and Traders Association, London

Van Ramshorst, H. and Thomas, P. C. (1988) Digestion in sheep of diets containing barley chemically treated to reduce its ruminal degradability. *Journal of the Science of Food and Agriculture*, **42**, 1–7

Wainman, F. W., Dewey, P. J. S. and Boyne, A. W. (1981) Compound feedstuffs for ruminants. *Third Report of the Feedstuffs Evaluation Unit*. Department of Agriculture and Fisheries for Scotland, Edinburgh

Webster, A. J. F., Dewhurst, R. J. and Waters, C. J. (1988) Alternative approaches to the characterisation of feedstuffs for ruminants. In *Recent Advances in Animal Nutrition – 1988*, (ed. W. Haresign and D. J. A. Cole). Butterworths, London. pp. 167–191

Whitelaw, F. G., Milne, J. S., Ørskov, E. R. and Smith, J. S. (1986) The nitrogen and energy metabolism of lactating cows given abomasal infusions of casein. *British Journal of Nutrition*, **55**, 537–556

Yates, F., Boyd, D. A. and Petit, G. H. N. (1942) Influence of changes in level of feeding on milk production. *Journal of Agricultural Science, Cambridge*, **32**, 428–456

Chapter 8

New techniques in the nutrition of grazing cattle

C. J. C. Phillips

Allowing cattle to harvest their own feed in the form of grazed crops is generally used in lower input systems than where crops are harvested by man and stored before feeding to cattle. Reductions in harvesting efficiency by cattle compared to man is partly or wholly offset by an absence of storage losses.

Advances in the nutrition of grazing cattle have not proceeded as rapidly as in some other fields of cattle production, mainly because of the stability and resistance to change of simple continuous grazing systems. Grazing systems tend to predominate where environmental conditions are suitable and lower inputs and outputs are required. As such, the need to keep inputs, in terms of labour, supplementary feed, fences etc. to a minimum are of paramount importance in grazing systems. There have been a number of new grazing methods derived for cattle which, while offering some theoretical benefits in terms of cattle nutrition, have faltered on practical/economic grounds. However, even if there have been delays in developing improved grazing systems for cattle, there is no doubt that knowledge of the grazing process has been greatly increased, which should eventually increase our ability to design improved and economic grazing systems that cater adequately for the nutrition and well-being of cattle.

The grazing process

Early research in the ingestion of grazed material by cattle described the periodicity of grazing (e.g. Hancock, 1950; Tayler, 1953), as well as suggesting factors that affect cattle grazing time. More recently, the main components of herbage ingestion – bite mass, biting rate and grazing time – have been studied. It is generally agreed that bite mass is the controlling factor regulating herbage intake, based on the circumstantial evidence that it shows a strong correlation with different sward conditions (Hodgson, 1981; Forbes and Hodgson, 1985), and grazing time and biting rate are modified in an attempt by the animal to maintain intake in adverse sward conditions. It has also been proposed that there is a ceiling to the maximum grazing time and biting rate (Jamieson and Hodgson, 1979; Phillips and Leaver, 1986a), the former imposed by the requirement of cattle for adequate rest (Wierenga, 1984) and the latter by the inability of cattle efficiently to prehend herbage at high biting rates.

The daily grazing period can be further subdivided into a number of grazing bouts (Phillips and Denne, 1988), the frequency of which appears to be largely

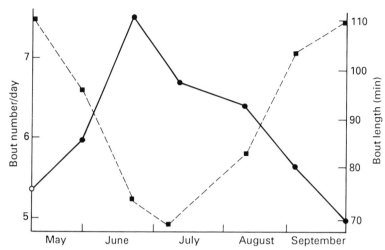

Figure 8.1 Variation in bout number (●—●) and bout length (■- - -■) over the grazing season. (From Hecheimi, Phillips and Murray-Evans, 1988)

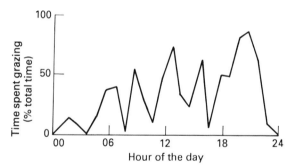

Figure 8.2 Periodicity of cattle grazing behaviour. (From Phillips and Leaver, 1986)

under the influence of seasonal factors (Hecheimi, Phillips and Murray-Evans, 1988). Although total daily grazing time is not affected by seasonal factors under constant sward conditions, the number of grazing bouts/day increases and the length of each bout decreases with increasing daylength (Figure 8.1). There is also a marked reluctance of cattle to eat at night (Phillips and Denne, 1988) even in hot climates (Alhassan and Kabuga, 1988), and the pattern of grazing observed (Figure 8.2) is probably a compromise between this factor and the need for ruminants to obtain a gradual input of feed over the day to stabilize rumen conditions and/or obtain adequate rest throughout the day. Under temperate conditions dairy cows normally exhibit four to six grazing bouts/day (mean 5.0 + 0.08) with a mean duration of 111 minutes each (Table 8.1). The night grazing bouts are more variable both in number and duration. The coefficient of variation (CV) for total daily grazing time is only 15%, suggesting that the cow can vary grazing time over the 24 h to buffer the effects at any one time of the day.

One of the main factors causing short-term variation in grazing behaviour is rainfall and, if this is prolonged so that it cannot be adequately buffered, grazing

Table 8.1 Dairy cow grazing bout number and length, and the coefficient of variation (CV) of bout number and length

	Day (07–23.00)	Night (23–07.00)
Bout no.	4.2	1.0
Bout length (min)	132	30
CV bout no. (%)	26	73
CV bout length (%)	34	77

From Phillips and Denne, 1988.

Table 8.2 Wet and dry herbage feeding behaviour

	Wet herbage	Dry herbage
Feeding time (h/day)	5.2	4.6
Rate of intake (gDM/min)	28.9	28.6
DM intake (kg/day)	9.1	7.0

Butris and Phillips, 1987.

DM: dry matter.

time is reduced by up to 1 h/day (Hinch, Thwaites and Lynch, 1982). Similar effects can be observed with cattle fed dry or soaked herbage where more accurate measurements of rate of dry matter (DM) intake can be made (Table 8.2). The reduction in DM intake with wet herbage results from reduced feeding time rather than rate of DM intake, suggesting that wet herbage has a lower rate of passage through the animal rather than there being a physical difficulty in prehending it. One possible reason for a lower rate of passage is reduced production of saliva which has been observed with wet compared with dry feeds. Early research by Campling and Balch (1961) showed that adding water to the rumen through a fistula has no effect on DM intake, confirming that it is not an effect of water *per se* in the rumen.

In hot climates high temperatures cause a reduction in daytime grazing but night-time grazing is increased in partial compensation.

Advances in grazing systems

Continuous grazing

Continuous grazing systems allow unrestricted grazing of the entire area throughout the grazing season, in comparison with rotational grazing systems which utilize intermittent grazing on a rotational basis.

One of the dominant influences on research into grazing systems for cattle has been the realization over the last 20 years that stocking rate is closely related to output/ha. Although theoretically stocking rate is curvilinearly related to output/ha, in practice the relationship is linear over the range of cattle stocking rates commonly used (Jones and Sandland, 1974). Doubt still exists as to whether production/animal is linearly or curvilinearly related to stocking rate (Mayne, Newberry and Woodcock, 1988), but the absence of any effect of stocking rate on milk production or herbage intake in the experiment of Baker and Leaver (1986) suggests a curvilinear relationship. More specifically, the relationship between herbage allowance and intake has been elucidated showing that, for dairy cows, DM intake begins to decline below approximately 40 g DM/kg liveweight, with only a slight increase in this value at higher milk yield levels (Figure 8.3).

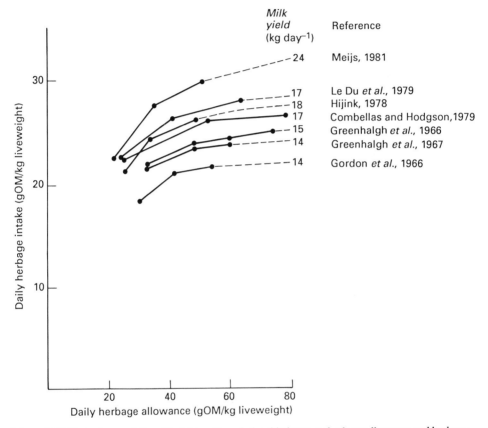

Figure 8.3 Effect of mean daily milk yield on the relationship between herbage allowance and herbage intake

A knowledge of this relationship is of limited use to farmers as they cannot easily measure herbage allowances, but the relationship between the major component of herbage allowance – herbage height – and DM intake has been evaluated (Le Du and Hutchinson, 1982), and practical recommendations are available for target herbage heights at different stages in the grazing season to maintain production/animal and /ha at optimum levels (Table 8.3). Additional information on the effects of supplements and sward type on target sward height may enable more detailed guidelines to be made available to farmers in the near future.

In recent years there has been a greater realization of the need to control spring growth by close grazing and the use of high stocking rates at this time aids farms

Table 8.3 Target sward heights (cm) for dairy cows

	Autumn-calving cows	Spring-calving cows
April–June	6–8	8–10
July–September	8–10	10–12

Courtesy of Mayne, 1988.

making large quantities of conserved forage at this period of rapid growth. At the end of the growing season farmers are now more aware of the deficiencies of autumn herbage. Intake of autumn herbage is typically between 11% (Corbett, Langlands and Reid, 1963) and 19% (Stehr and Kirchgessner, 1976) less than spring herbage. There are many factors responsible for this, but most important are the high dead matter content of the sward and the faecal contamination and rejection of the sward. Herbage dead matter content increases from approximately 10% in spring to 40% in autumn (Le Du, Baker and Newberry, 1981) and by the end of the grazing season 20–30% of the sward is rejected due to faecal contamination (Phillips and Leaver, 1985a,b). The increase in dead matter content in the lower horizons of the sward has the effect of reducing the grazed horizon, which together with the reduction in grazed area due to faecal rejection, emphasizes the need for a higher target sward height in autumn than spring. Attempts to overcome the rejection of herbage around dung pats by conditioning cows (MAFF, 1968; Garstang and Mudd, 1971) or dispersing the faeces with a chain harrow (Weeda, 1967; MAFF, 1969; Reid, Greenhalgh and Aitken, 1972) have not been successful. Some success has been achieved by spraying rejected areas with molasses (Marten and Donker, 1964), or alternatively topping the rejected herbage will hasten the return of these areas into the grazed sward.

Research has shown that even at the same digestibilities, autumn herbage produces lower liveweight gains in beef cattle than spring herbage (Lonsdale and Tayler, 1971), probably because it contains more structural and less water soluble carbohydrate and rumen retention time is greater. Milk yields of dairy cows have not been found to be reduced with cut autumn herbage compared with cut spring herbage, possibly because of the higher acetate:propionate ratio from autumn herbage (Bath and Rook, 1961), which stimulates a diversion of nutrients to milk production.

Research on the integration of conservation and grazing has produced a valuable management plan for continuously grazed cattle (Ilius, Lowman and Hunter, 1987). This plan shows that setting aside 25–30% of the grazed area at the start of the grazing season, to be conserved if not required for grazing, will facilitate intensive grassland management at high levels of nitrogen fertilizer application. Included in the management plan are decision rules for determining when to allow access to the set-aside area.

Rotational grazing

Although under intensive grazing conditions with single species swards no difference has been found in animal output between rotational and continuous grazing systems (Figure 8.4) (Ernst, Le Du and Carlier, 1980), this is not the case

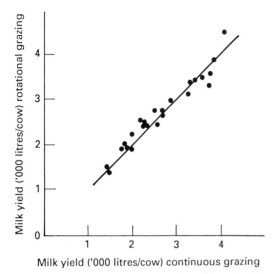

Figure 8.4 Milk production levels under rotational and continuous grazing. (Courtesy of Le Du and Hutchinson, 1982)

for less intensively managed swards. Multispecies swards can suffer under continuous grazing as the preferred species may be subject to an increased frequency of defoliation that depletes the contribution of that species to the sward. The use of rotational grazing allows plants that are not tolerant to continuous grazing to recover, e.g. the clover content of a grass:clover sward will be increased under rotational grazing. Also important in protecting the preferred species in a rotational grazing system is prevention of overgrazing (Eckert and Spencer, 1987). Residual sward height is therefore the most important target for the satisfactory management of rotational grazing. Considerable interest is now shown in employing rotational grazing systems for rangeland cattle grazing (Heitschmidt, Dowhower and Walker, 1987). An additional advantage of rotational grazing under rangeland conditions is reduced energy expenditure because of the greater concentration of forage and hence reduced grazing time and increased rate of ingestion (Table 8.4).

A specific form of rotational grazing that potentially offers some nutritional advantages is the leader-follower system, where stock with high intake

Table 8.4 Grazing behaviour of steers under continuous and rotational grazing on crested wheatgrass rangeland

	Continuous	Rotational
Grazing time (h/day)	10.9	9.7
Ingestion rate (gOM/min)	5.4	10.3
OM intake (kg/day)	3.5	6.0

Courtesy of Olson and Malechek, 1988.

OM: organic matter.

requirements are grazed ahead of stock with low intake requirements. With lactating cows, as we saw earlier, there is very little effect of yield level on the relationship between herbage intake and allowance and it has also been found that high yielders are unable to graze for longer despite their higher nutrient demands (Phillips and Leaver, 1986b). Conceivably both high and low yielders are at the limit of their grazing intake potential, and with dairy cows a leader-follower system has not been found markedly to increase milk yields (Archibald, Campling and Holmes, 1975; Mayne, Newberry and Woodcock, 1988). An exception to this could be during spring growth of herbage when a significant relationship between milk yield and grazing time has been found (Phillips, 1983) and where good grazing conditions can supply the additional nutrient requirements of high yielders without resulting in excessively high grazing times. It has been suggested that a system where dry cows, with low nutrient demands, follow early lactation cows would be the most appropriate leader-follower system for dairy herds (Archibald, Campling and Holmes, 1975). Good results have been obtained with a leader-follower system for growing cattle (Leaver, 1970). In this system calves are grazed ahead of older heifers and, because they are more selective in their grazing and also susceptible to worm infections, increases in liveweight gain/ha of approximately 20% have been recorded.

Supplementation of grazing cattle

Forages

One of the greatest limitations to cattle production from grazed temperate swards is the understocking necessitated by the high variation in level of available herbage from grazed swards. Some variation in herbage availability can be predicted, e.g. a spring flush of grass, and wastage of herbage minimized by having a higher stocking rate at this time. However, further variation in herbage availability arises from unpredictable variation in rainfall and solar radiation levels. Variation in feed availability can be eliminated by supplementing pasture with conserved forage. When coupled with an increase in stocking rate to increase the level of utilization of grazed herbage, this policy can result in an increase in the efficiency of grassland utilization on the farm. However, most forms of conserved forage support lower levels of animal production than fresh grazed herbage, and if stocking rate is increased too much, forcing the cattle to rely too heavily on lower quality conserved material, animal production levels will fall. Alternatively, if stocking rates are not increased herbage utilization levels will fall.

The amount of conserved forage required will vary with the CV of annual herbage production and the increase in stocking rate adopted. As the annual rainfall decreases the CV of annual rainfall increases, and in arid regions the annual CV of herbage production averages 34% (Le Houerou, Bingham and Skerbeck, 1988) compared with only 23% in temperate regions (Garwood, Tyson and Clement, 1977).

Forage quality is important in maintaining the correct buffering capacity of the grazed sward. If forage quality is greater than the grazed sward, animal production will increase, but management of the grazed sward will be difficult if cattle prefer the forage to the grazed sward. If forage quality is too low compared with the grazed sward, e.g. straw, only small benefits in terms of animal production will result, particularly if nutrient demands are high. The ideal forage supplement is not eaten in preference to the grazed sward, but complements it well nutritionally and is able to restore the intake of nutrients to the optimum level during periods of restricted herbage availability. Some specific nutrient deficiencies occur on pastures when they are the sole feed, e.g. fibre on very young pasture, rumen-undegraded protein particularly on mature tropical pastures and micronutrients where the soils are deficient. In the case of young herbage, attempts to restore milk fat content with supplementary silage or hay have been successful in some cases (Walsh, 1969; Phillips and Leaver, 1985b), but because the mature forage has a lower energy and protein content than young herbage, milk yield and milk protein content are generally reduced. Where adequate young herbage is available, only small quantities of high fibre forage will be consumed. The use of conserved forage to buffer reduced pasture intake in hot climates because of high daytime temperatures has not been found to be successful (Seath and Miller, 1947; Seath et al., 1956).

Greater responses to supplementary forage are obtained from high yielding than low yielding cows (Phillips and Leaver, 1985a), in contrast to the response to higher herbage availability. The reason for this is probably because conserved forage is ingested at two to three times the rate of grazed pasture (Phillips, 1988a) and can therefore increase the DM intake of high yielders more effectively than additional pasture where behavioural limitations appear to limit intake.

The effects on whole farm land utilization efficiency depend on the type of forage fed and the relative efficiency with which pasture is grazed and forage conserved. Ensiled herbage produces higher DM yields than grazed pasture and the relative losses from the two forms of harvesting determine whether overall efficiency is increased (Phillips, 1988b). Conserving and feeding forage requires more support energy, mechanization and labour than grazing pasture (White, Wilkinson and Wilkins, 1983), and the availability of these determines the attractiveness of buffer feeding. Paradoxically the areas most likely to benefit from buffer feeding, i.e. the arid regions where the CV of annual herbage production is high, are those areas where it is difficult and expensive to conserve forage.

The buffer feeding system can result in an increased efficiency of pasture utilization by increasing farmer confidence to raise stocking rates and reduce herbage rejection. Too great a reliance on low quality conserved forage during the grazing season can however reduce both milk production and the reproductive performance of high yielding dairy cows (Phillips, 1988b).

Concentrates

Reviews (Leaver, Campling and Holmes, 1968; Jennings and Holmes, 1985) of experimental feeding of supplementary concentrates to dairy cows on temperate and tropical pastures produced respectively mean response rates of 0.41 and 0.63 kg milk/kg concentrate. The variation in response rate was very high with CVs of 50 and 63% respectively. Recent research has attempted to highlight those situations where high response rates will be obtained. The main sources of variation in response rate are the quality and quantity of pasture, the quality of supplement and the stage of lactation of the cows. With regard to the quality of pasture it is clear that tropical pastures, which are of lower digestibility than temperature pastures, produce a greater response rate. At low herbage availabilities response rate is also increased and substitution rate decreased. Perhaps the most dramatic illustration of the latter is the work of Meijs and Hoekstra (1984) who found a substitution rate of

0.1 kg herbage organic matter (OM)/kg concentrate OM at 16 kg OM allowance/cow and 0.5 kg OM/kg concentrate OM at 24 kg OM allowance/cow. They concluded that substitution rates depended largely on the level of herbage availability and probably the difference between the nutrient requirement and nutrient supply from herbage.

With regard to concentrate composition, Meijs (1986) has shown with dairy cows grazing temperate pasture that concentrates based on digestible fibre will produce greater yields of milk and, in particular, milk fat than starch-based concentrates, which tend to encourage lipogenesis in accordance with glucogenic theory. The most likely explanation for this is that starchy concentrates decrease rumen pH and reduce fibre digestion, resulting also in lower forage intakes. The protein content of the concentrate is important first in supplementing tropical pastures where DM and crude protein (CP) intakes tend to be low (Kaiser and Ashwood, 1982a), and secondly where high levels of concentrates are fed stimulating milk production and protein requirements (Kaiser and Ashwood, 1982b). On temperate pastures no effect of supplementary urea has been found, although inclusion of fishmeal in concentrates has increased the response rate from 1.2 to 1.6 kg milk/kg concentrate in high yielding cows (Le Du and Newberry, 1980). Supplementation of tropical pasture with protected protein has shown response rates of 2.2 (Rogers and Porter, 1978) to 2.4 kg milk/kg concentrates (Stobbs, Minson and MacLeod, 1977), the response being greater when the energy requirements of the cows are satisfied (Davison, Marschke and Brown, 1982). Inclusion of sodium bicarbonate in concentrates has been shown to alleviate up to 40% of the depression in milk fat content caused by grazing cows on low fibre herbage (Kaiser and Ashwood, 1982b).

The response to supplementary concentrates is greater with high yielding cows (Laird and Walker-Love, 1962) as was found with supplementary forage, but few experiments have been conducted at yields in excess of 25 kg/cow per day. The residual effect of concentrate supplements is large if the level of nutrition after supplementation is inadequate (Burstedt, 1981). On tropical pastures the average direct response is 0.4 kg milk/kg concentrates, whereas the average whole lactation response is 0.8 kg milk/kg concentrates (Jennings and Holmes, 1985).

For growing cattle it is often less economic to supplement pasture than for dairy cows because of the lower value of the end product, even though dairy cattle are able to buffer an intake restriction by mobilizing body tissue, whereas all the effects of intake restriction for growing cattle are transferred directly to loss of output. It is still necessary to supplement pastures in areas of the world where variation in pasture availability is high, and the traditional methods of achieving this have involved feeding supplementary rumen-degradable nitrogen in the form of urea and energy in the form of molasses, cereals etc. Other more fibrous products that can be used to supplement tropical pasture include bagasse, hay and silage and, more recently, byproducts such as ammonia-treated crop residues, and poultry wastes (Brown *et al.*, 1987) have been found to be useful suppliers of rumen-degradable nitrogen (Brown *et al.*, 1987).

Minerals and rumen modifiers

In addition to the common mineral deficiencies of copper, cobalt etc., it has recently been established that grazing cattle can be deficient in chromium, tin, vanadium and nickel, and also that modern industrial practices have increased the possibilities of cadmium, mercury, lead and arsenic toxicity or adverse effects of these elements on the metabolism of essential minerals, particularly copper (Williams and McDowell, 1986).

Conventional single dose methods of administering minerals or rumen modifiers such as sodium monensin suffer from the need for repeated application and excessive concentration of drug release after dosing (Allen and Drake, 1986). Controlled release systems can now supply therapeutic agents either continuously or in pulses and are based on diffusion through an insoluble polymer or glass. Controlled release glass contains cations, e.g. calcium, which determine the solubility of the glass and anions, e.g. phosphate. Controlled release glass administrators are usually given orally and have successfully released copper, cobalt and zinc at constant rates for up to one year (Allen and Drake, 1986). Rumen modifiers have been successfully delivered from a polymer-based device over 150 days (Lawrence *et al.*, 1986), thus rendering it possible to administer monensin throughout the year in beef-based systems relying on summer grazing.

Legume pasture grazing

Interest in legume-based pastures has received a fresh impetus in many developed countries as a means to maintain profitability by input reduction rather than increased output are being sought.

The ability of legumes partially to substitute for fertilizer nitrogen and hence maintain the carrying capacity of the sward at approximately 70–80% (Steen and Laidlaw, 1987) of the capacity of a sward receiving 360 kg nitrogen/ha is well established, as is the improvement in the growth rates of cattle grazing legume pastures compared with grass-based pastures. Recently the milk production of dairy cows grazing white clover or perennial ryegrass has been compared (Table 8.5), showing that clover increases milk yield and protein content and decreases fat

Table 8.5 Milk production from white clover and perennial ryegrass

	White clover	Perennial ryegrass
Milk yield (kg/day)	25.0	22.2
Content (g/kg) of:		
Fat	38.7	42.2
Protein	31.8	30.3

Courtesy of Thompson, *et al.*, 1985.

content. The maintenance of high white clover contents in temperate grass-based swards requires short sward height and minimum fertilizer nitrogen application (Frame and Boyd, 1987; Fulkerson and Mitchell, 1987; Murray-Evans, Phillips and Hecheimi, 1988). Murray-Evans *et al.* (1988) found that the clover content of the sward averaged 32% at 4 cm and 22% at 8 cm, but at 4 cm milk production was considerably reduced unless a silage supplement was fed (Table 8.6). A persistent problem with grazing high legume swards is bloat. Some advances have been made in identifying that it is probably differences in rumen fluid volume between cattle that determine bloat susceptibility (Cockrem *et al.*, 1987) and that feeding

Table 8.6 Milk yield and bloat incidence from high clover swards with and without a forage supplement

	4 cm	*4 cm + silage*	*8 cm*	*8 cm + silage*
Milk yield (kg/day)	18.9	21.1	22.5	22.7
Bloat incidence/cow/week	0.17	0.08	0.17	0.07

Murray-Evans, Hecheimi and Phillips, 1988.

supplementary forage reduces the incidence of bloat (Murray-Evans, Phillips and Hecheimi, 1988). Preventative measures such as breeding bloat resistant cows or clover, or feeding supplementary forage, require further investigation before satisfactory control of bloat can be achieved.

Lucerne is an alternative legume to clover favoured in drought susceptible areas. This has higher production levels than grass:clover pastures and produces fewer bloat problems (Douglas, 1986). Energy content of the herbage can however be low and a strict rotational grazing system is also needed.

Conclusions

Our knowledge of the grazing process by cattle has advanced significantly in recent years, and much of this knowledge will be of use in devising systems of grazing management that can overcome the difficulties caused by predictable and unpredictable variation in herbage production levels. This is particularly true in arid areas where there is high variation in productivity levels and herbage density is often low, both tending to mitigate against conventional continuous grazing systems. In temperate areas there is an inherent stability in continuous grazing systems but the potential to achieve high milk yields by strategically supplementing ryegrass swards exists. Our ability to overcome micronutrient deficiencies using slow releasing intraruminal devices is greatly enhanced and should bring major rewards, particularly for extensively grazed stock.

Research in legume utilization has received renewed impetus in developed countries where surpluses of cattle products have encouraged producers to seek methods of reducing inputs. Much is now known about grazing methods to ensure high production levels from pasture and cattle, but the problem of bloat control remains as yet unsolved.

References

Alhassan, W. S. and Kabuga, J. G. (1988) Dry and wet seasonal studies on the grazing behaviour of native N'dama and imported Friesian bulls in a humid tropical environment. *Applied Animal Ethology*, **19**, 245–254

Allen, W. M. and Drake, C. F. (1986) Controlled release of mineral nutrients and other pharmaceuticals for sheep and cattle. In *Proceedings of the Sixth International Conference on Production Disease in Farm Animals*, Belfast, September, 1986, pp. 171–174

Archibald, K. A. E., Campling, R. C. and Holmes, W. (1975) Milk production and herbage intake of dairy cows kept on a leader and follower grazing system. *Animal Production,* **21**, 147–156

Baker, A. M. C. and Leaver, J. D. (1986) Effect of stocking rate in early season on dairy cow performance and sward characteristics. *Grass and Forage Science*, **41**, 333–340

Bath, I. H. and Rook, J. A. F. (1961) The effect of stage of growth of S23 perennial ryegrass on the production of volatile fatty acids in the rumen of the cow. *Proceedings of the Nutrition Society*, **20**, 15 (abstract)

Brown, C. J., Loe, W. C., Parham, R. W. and Johnson, Z. B. (1987) Heifer development and reproduction. *Bulletin 909*, Arkansas Agricultural Experiment Station, USA

Burstedt, E. (1981) The effect of supplementary feed on pasture utilization and milk yield from high yielding dairy cows. In *Proceedings of the 32nd meeting of the European Association for Animal Production*, **IV** -2, 1–7

Butris, G. Y. and Phillips, C. J. C. (1987) The effect of herbage surface water and the provision of supplementary forage on the intake and feeding behaviour of cattle. *Grass and Forage Science*, **42**, 259–264

Campling, R. C. and Balch, C. C. (1961) Factors affecting the voluntary intake of food by cows 1. Preliminary observations on the effect, on the voluntary intake of hay, of changes in the amount of reticulo-ruminal contents. *British Journal of Nutrition*, **15**, 523–530

Cockrem, F. R. M., McIntosh, J. T., McLaren, R. D. and Morris, C. A. (1987) The relationship between volume of rumen contents and genetic susceptibility to pasture bloat in cattle. *Animal Production*, **45**, 43–48

Combellas, T. and Hodgson, J. (1979) Herbage intake and milk production by grazing dairy cows 1. The effect of variation in herbage mass and daily herbage allowance in a short term trial. *Grass and Forage Science*, **34**, 209–214

Corbett, J. L., Langlands, J. P. and Reid, G. W. (1963) Effects of season of growth and digestibility of herbage on intake by grazing dairy cows. *Animal Production*, **5**, 119–129

Davison, T. M., Marschke, R. J. and Brown, G. W. (1982) Milk yields from feeding maize silage and meat-and-bone meal to Friesian cows grazing a tropical grass and legume pasture. *Australian Journal of Experimental Agriculture and Animal Husbandry*, **22**, 147–154

Douglas, J. A. (1986) The production and utilization of lucerne in New Zealand (review paper). *Grass and Forage Science*, **41**, 81–128

Eckert, R. E. and Spencer, J. S. (1987) Diet and forage intake of cattle on desert grassland range. *Journal of Range Management*, **40**, 156–159

Ernst, P., Le Du, Y. L. P. and Carlier, L. (1980) Animal and sward production under rotational and continuous grazing management – a critical appraisal. In *The Role of Nitrogen in Intensive Grassland Production*, (ed. W. H. Prins and G. M. Arnold). Centre for Agricultural Publishing and Documentation, Wageningen, The Netherlands, pp. 119–126

Forbes, T. D. A. and Hodgson, J. (1985) Comparative studies of the influence of sward conditions on the ingestive behaviour of cows and sheep. *Grass and Forage Science*, **40**, 69–77

Frame, J. and Boyd, A. G. (1987) The effect of strategic use of fertilizer nitrogen in spring and/or autumn on the productivity of a perennial ryegrass/white clover sward. *Grass and Forage Science*, **42**, 429–438

Fulkerson, W. J. and Mitchell, P. J. (1987) The effect of height and frequency of mowing on the yield and composition of perennial ryegrass-white clover swards in the autumn to spring periods. *Grass and Forage Science*, **42**, 169–174

Garstang, J. R. and Mudd, C. H. (1971) The rejection of contaminated herbage by dairy cows. *Journal of the British Grassland Society*, **26**, 194 (abstract)

Garwood, E. A., Tyson, K. C. and Clement, C. R. (1977) A comparison of yield and soil conditions during 20 years of grazed grass and arable cropping. *Technical Report no. 21*, Grassland Research Institute, Hurley, UK

Gordon, C. H., Derbyshire, J. C., Alexander, C. W. and McCloud, D. E. (1966) Effects of grazing pressure on the performance of dairy cattle and pastures. In *Proceedings of the X International Grassland Congress*, Helsinki, July 1966 (ed. A. G. G. Hill, V. U. Mustonen, S. Pulli and M. Latvala), pp. 470–479

Greenhalgh, J. F. D., Reid, G. W., Aitken, J. N. and Florence, E. (1966) The effects of grazing intensity on herbage consumption and animal production 1. Short-term effects in strip-grazed dairy cows. *Journal of Agricultural Science, Cambridge*, **67**, 13–23

Greenhalgh, J. F. D., Reid, G. W. and Aitken, J. N. (1967) The effects of grazing intensity on herbage consumption and animal production 2. Long-term effects in strip-grazed dairy cows. *Journal of Agricultural Science, Cambridge,* **69**, 217–223

Hancock, J. (1950) Studies in monozygotic cattle twins IV uniformity trials: grazing behaviour. *New Zealand Journal of Science and Technology,* **32**, 22–59

Heicheimi, K., Phillips, C. J. C. and Murray-Evans, J. P. (1988) Variation in the ingestive behaviour of dairy cows grazing a high clover pasture with or without a silage buffer. In *Proceedings of Research Meeting no. 1.* British Grassland Society, Hurley

Heitschmidt, R. K., Dowhower, S. L. and Walker, J. W. (1987) 14-vs.42-paddock rotational grazing: forage quality. *Journal of Range Management,* **40**, 315–317

Hijink, J. W. F. (1978) Snijmais bijvoeren aan koeien inde weideperiode (Supplementary feeding of maize silage to dairy cows during the grazing season). *Publication no. 12.* Proefstation voor de Rundveehouderij, Lelystad

Hinch, G. N., Thwaites, C. J. and Lynch, J. J. (1982) A note on the grazing behaviour of young bulls and steers. *Animal Production,* **35**, 289–291

Hodgson, J. (1981) Variations in the surface characteristics of the sward and the short-term rate of herbage intake in calves and lambs. *Grass and Forage Science,* **36**, 49–58

Ilius, A. W., Lowman, B. G. and Hunter, E. A. (1987) Control of sward conditions and apparent utilization of energy in the buffer grazing system. *Grass and Forage Science,* **42**, 283–296

Jamieson, W. S. and Hodgson, J. (1979) The effect of daily herbage allowance and sward characteristics upon the ingestive behaviour and herbage intake of calves under strip-grazing management. *Grass and Forage Science,* **34**, 273–282

Jennings, P. G. and Holmes, W. (1985) Supplementary feeding to dairy cows grazing tropical pasture: a review. *Tropical Agriculture,* **62**, 266–272

Jones, R. L. and Sandland, R. L. (1974) The relation between animal gain and stocking rate. Derivation from the results of grazing trials. *Journal of Agricultural Science, Cambridge,* **83**, 335–342

Kaiser, A. G. and Ashwood, A. M. (1982a) Response by grazing cattle to supplementary protein. In *New South Wales Agricultural Research Station Biennial Report,* pp. 27–28

Kaiser, A. G. and Ashwood, A. M. (1982b) Influence of wheat, soyabean and sodium bicarbonate supplements on milk production and composition. In *New South Wales Agricultural Research Station Biennial Report,* pp. 29–30

Laird, R. and Walker-Love, J. (1962) Supplementing high-yielding cows at pasture with concentrates fed at a level determined by milk yield and season. *Journal of Agricultural Science, Cambridge,* **59**, 233

Lawrence, K., Davis, C., Dobson, M. B. and Buchanan, A. G. (1986) Evaluation of monensin sodium from a ruminal delivery device as a growth promoter for pastured fattening cattle. In *Proceedings of the Sixth International Conference on Production Disease in Farm Animals,* Belfast, September 1986, pp. 277–280

Le Du, Y. L. P., Baker, R. D. and Newberry, R. D. (1981) Herbage intake and milk production of grazing dairy cows. 3. The effect of grazing severity under continuous stocking. *Grass and Forage Science,* **36**, 307–318

Le Du, Y. L. P., Combellas, J., Hodgson, J. and Baker, R. D. (1979) Herbage intake and milk production by grazing dairy cows. 2. The effect of level of winter feeding and level of herbage allowance. *Grass and Forage Science,* **34**, 249–260

Le Du, Y. L. P. and Hutchinson, M. (1982) Grazing. In *Milk From Grass,* (ed. C. Thomas and J. W. O. Young), ICI, Billingham; Grassland Research Institute, Hurley, pp. 43–47

Le Du, Y. L. P. and Newberry, R. D. (1980) The milk production of grazing cows. In *Grassland Research Institute Annual Report,* pp. 86–88

Le Houerou, H. N., Bingham, R. L. and Skerbeck, W. (1988) Relationship between the variability of primary production and the variability of annual precipitation in world arid lands. *Journal of Arid Environments,* **15**, 1–18

Leaver, J. D. (1970) A comparison of grazing systems for dairy herd replacements. *Journal of Agricultural Science, Cambridge,* **75**, 265–272

Leaver, J. D., Campling, R. C. and Holmes, W. (1968) Use of supplementary feeds for grazing dairy cows. *Dairy Science Abstracts,* **30**, 355–361

Lonsdale, C. R. and Tayler, J. C. (1971) The effect of season of harvest and of milling on the nutritive value of dried grass. *Animal Production*, **13**, 384 (abstract)

MAFF (1968) Rejection of grass by cows. In *Great House Experimental Husbandry Farm Annual Report*, Ministry of Agriculture, Fisheries and Food UK, pp. 28–29

MAFF (1969) Pasture contamination. In *Rosemaund Experimental Husbandry Farm Annual Report*, Ministry of Agriculture, Fisheries and Food UK, pp. 28–30

Marten, G. C. and Donker, J. D. (1964) Selective grazing induced by animal excreta. *Journal of Dairy Science*, **47**, 773–776

Mayne, C. S. (1988) Grazing management for the dairy herd. *Agriculture in Northern Ireland*, **2**, (7), 10–11

Mayne, C. S., Newberry, R. D. and Woodcock, S. C. F. (1988) The effects of a flexible grazing management strategy and leader/follower grazing on the milk production of grazing dairy cows and on sward characteristics. *Grass and Forage Science*, **43**, 137–150

Meijs, J. A. C. (1981) Het effect van kracktvoerbijvoedering op de graspname van weidende melkkoeien 1. Verslag van de grasepnameproef in 1981. (The influence of concentrate supplementation on herbage intake by grazing dairy cows. 1. Report of the experiment on herbage intake in 1981). *Report no. 143*, Institut voor Veevorderingsonderzoek 'Hoorn', Lelystad, The Netherlands

Meijs, J. A. C. (1986) Concentrate supplementation of grazing dairy cows 2. Effect of concentrate composition on herbage intake and milk production. *Grass and Forage Science*, **41**, 229–236

Meijs, J. A. C. and Hoekstra, J. A. (1984) Concentrate supplementation of grazing dairy cows. 1. Effect of concentrate intake and herbage allowance on concentrate intake. *Grass and Forage Science*, **39**, 59–66

Murray-Evans, J. P., Hecheimi, K. and Phillips, C. J. C. (1988) The effect of herbage height and the provision of high and low-quality silage for dairy cows grazing a high clover pasture on sward composition, milk production and ingestive behaviour. *Animal Production*, **46**, 488 (abstract)

Olson, K. C. and Malechek, J. C. (1988) Heifer nutrition and growth on short duration grazed crested wheatgrass. *Journal of Range Management*, **41**, 259–263

Phillips, C. J. C. (1983) Conserved forage as a buffer feed for dairy cows. *PhD thesis*, University of Glasgow, UK

Phillips, C. J. C. (1988a) The use of conserved forage as a supplement for grazing dairy cows. *Grass and Forage Science*, **43**, 215–230

Phillips, C. J. C. (1988b) Partial storage-feeding for dairy cows. *Research and Development in Agriculture*, (in press)

Phillips, C. J. C. and Denne, S. P. J. (1988) Variation in the grazing behaviour of dairy cows, measured by a Vibrarecorder and bite count monitor. *Applied Animal Ethology*, **21**, 329–335

Phillips, C. J. C. and Leaver, J. D. (1985a) Supplementary feeding of forage to dairy cows 1. Offering hay to cows at high and low stocking rates. *Grass and Forage Science*, **40**, 183–192

Phillips, C. J. C. and Leaver, J. D. (1985b) Supplementary feeding of forage to dairy cows 2. Offering silage in early and late season. *Grass and Forage Science*, **40**, 193–199

Phillips, C. J. C. and Leaver, J. D. (1986a) Seasonal and diurnal variation in the grazing behaviour of dairy cows. In *Proceedings of the British Grassland Society Symposium* no. 19, 1985, British Grassland Society, Hurley, pp. 98–104

Phillips, C. J. C. and Leaver, J. D. (1986b) The effect of forage supplementation on the grazing behaviour of dairy cows. *Applied Animal Behavioural Science*, **16**, 233–247

Reid, G. W., Greenhalgh, J. F. D. and Aitken, J. N. (1972) The effects of grazing intensity on herbage consumption and animal production. *Journal of Agricultural Science, Cambridge*, **78**, 491–496

Rogers, G. and Porter, R. (1978) Wilted silage as a supplement for pasture fed cows. In *Annual Report, Ellinbank Dairy Research Institute*, Victoria Department of Agriculture, Australia, pp. 57–58

Seath, D. M., Lassiter, C. A., Davies, C. L., Rust, J. W. *et al.* (1956) The supplemental value of alfalfa hay when fed to cows on pasture. *Journal of Dairy Science*, **39**, 274–279

Seath, D. M. and Miller, G. D. (1947) Effect of hay feeding in summer on milk production and grazing performance of dairy cows. *Journal of Dairy Science*, **30**, 921–926

Steen, R. W. J. and Laidlaw, A. S. (1987) The role of white clover in grassland for beef production. In *Annual Report of the Agricultural Research Institute of Northern Ireland, 1986–87*, pp. 14–22

Stobbs, T. H., Minson, D. J. and Macleod, M. M. (1977) The response of dairy cows grazing a nitrogen fertilized grass pasture to a supplement of protected casein. *Journal of Agricultural Science, Cambridge,* **89**, 137–141

Stehr, W. and Kirchgessner, M. (1976) The relationship between the intake of herbage grazed by dairy cows and its digestibility. *Animal Feed Science and Technology,* **1**, 53–60

Tayler, J. C. (1953) The grazing behaviour of bullocks under two methods of management. *British Journal of Animal Behaviour,* **1**, 72–77

Thompson, D. J., Bewer, D. E., Haines, M. J., Cammell, S. B. *et al.* (1985) Yield and composition of milk from Friesian cows grazing either perennial ryegrass or white clover in early lactation. *Journal of Dairy Research,* **52**, 17–31

Walsh, J. P. (1969) Effect of feeding on composition of cows' milk in late winter–early spring. *Irish Journal of Agricultural Research,* **8**, 319–327

Weeda, W. C. (1967) The effect of cattle dung patches on pasture growth, botanical composition and pasture utilization. *New Zealand Journal of Agricultural Research,* **10**, 150–159

White, D. J., Wilkinson, J. M. and Wilkins, J. (1983) Support energy use in animal production. In *Efficient Grassland Farming – 1982,* (ed. A. J. Corrall). British Grassland Society, Hurley, pp. 33–42

Wierenga, H. K. (1984) The social behaviour of dairy cows: some differences between pasture and cubicle systems. In *Proceedings of the International Congress on Applied Ethology in Farm Animals – 1984,* (ed. J. Unshelm, G. van Putten and K. Zeeb), KTBL, Kiel, Germany, pp. 135–138

Williams, S. N. and McDowell, L. R. (1986) Newly discovered and toxic elements. In *Nutrition of Grazing Ruminants in Warm Climates,* (ed. L. R. McDowell). Academic Press, London, pp. 317–338

Chapter 9

New techniques in environmental control for cattle

I. F. M. Marai and J. M. Forbes

Cattle inhabit almost all areas of the world, with the exception of areas of extreme cold and heat where mammals cannot generally survive. In almost all other areas cattle can be and are profitably kept for meat and milk production with only a minimum level of environmental control. Recently though there has been much interest in ameliorating the effects of the environment that reduce cattle production levels below maximum for the nutritional and genetic conditions operating. Cattle are relatively cold-tolerant by virtue of their size and coat characteristics and cold stress may be simply overcome by the provision of shelter to conserve the heat produced by the animal. The scientific knowledge in this field has progressed to the state of detailed modelling (*see* Chapter 14). Heat stress is more difficult to overcome, even though cattle have a well developed evaporative heat mechanism, particularly since many of the countries in which it occurs have severe financial constraints in the provision of elaborate control measures.

The seasonal nature of breeding in ruminant species is well documented and is currently the subject of a great deal of research, both to elucidate the mechanisms and to develop methods for out-of-season breeding. The annual rhythm of reproduction in these species is driven by the regular pattern of change of photoperiod, but it is only recently that photoperiod has been established as affecting growth and lactation directly in cattle.

Alleviation of heat stress

Physical techniques

Shade

The alleviation of the radiant heat load on animals may be achieved by the provision of shade in the form of a roof for cattle fed conserved feeds, and trees for grazing cattle. An overhead shade should be positioned with its length orientated north–south and should provide a solid area of shade between 3–4 m high and 5–8 m wide. Shades orientated east–west provide a slightly cooler environment because sun exposure to any part of the sheltered area is limited, but this tends to cause problems with excessive accumulation of wet manure. The shaded area should be located away from the feed area because, first, cattle should be encouraged to feed during the cool night and a shade effectively insulates the area underneath it from radiant heat loss, and second, the area will be less muddy

(Wiersma *et al.*, 1984). The shelter used should have high insulating properties and a low radiation coefficient (Ansell, 1976); plant material such as palm leaves, straw or grass are suitable except that they can harbour vermin. Manufactured roofing sheets of asbestos or metal are often used and can be made less absorbent by painting with a reflective paint or by using two sheets, one on top of the other with an air gap in between. In extreme environments, however, some additional heat stress will remain (Igono *et al.*, 1985) and must be counteracted by physical, physiological or nutritional techniques.

Air conditioning

Air conditioning is the most effective method of controlling heat stress in cattle, particularly in humid areas, but the operating costs are usually prohibitive because of the large amount of water that must be removed from the air. Evaporative coolers using pre-cooled air usually need to supply at least $0.5\,m^3/s$ of air flow per cow (2–2.5 air changes/min) in the target area (Wiersma *et al.*, 1984).

A new and more efficient method of supplying cool air to reduce body temperature is by the provision of cool air breathing hoods. This avoids the need to cool large quantities of air of which only a small proportion is used to cool the animals (Ansell, 1976). Another alternative to complete air conditioning is the use of desert coolers which rely on a forced draught over a wetted surface to cool the air, or individual forced air jets supplied to each cubicle (Igono *et al.*, 1985).

Sprinklers

Sprinklers, usually operating intermittently during the hottest part of the day, may be simply and cheaply installed to provide the cattle with an artificial sweating mechanism. As it is the latent heat of evaporation that is utilized the temperature of the water is immaterial (Ansell, 1976). The use of sprinklers adds about 40 litres water/h to the floor surface and good drainage is therefore essential (Igono *et al.*, 1985). It should be noted that when the environmental temperature is equal to or greater than body temperature, heat dissipation methods using sprinkling will be more effective than air conditioning (unless the air is cooled). Coupling sprinklers with enforced draughts from fans will increase the rate of evaporative heat loss (Wiersma *et al.*, 1984).

Shearing

Where sufficient protection exists from radiant heat in the form of shelters, shearing can usually increase the rate of evaporative heat loss in heat stressed cattle (Yeates, Edey and Hill, 1975). Such conditions are generally only found in intensively housed stock (Bianca, 1959).

Chilled drinking water

The beneficial effect of drinking cool water in reduction of the heat load is due to the heat dissipated by conduction as a result of the difference between the water and urine temperature. Moreover, the extra body water arising from the increase in water consumption helps dissipation of heat by increasing evaporative heat loss through sweating and respiration. The technique has also been found to increase roughage intake (Kamal, Kotby and El-Fouly, 1972), which usually is reduced in heat stress because of the additional stress caused by the heat increment of digestion.

Physiological techniques

Lactating cattle in particular are affected by heat stress because high yields of milk are associated with a high metabolic rate. As a result, milk production and tolerance to heat stress are likely to be inversely related (Smith and Mathewman, 1986). Heat stress depresses thyroid function and growth hormone levels, but the former can be restored by feeding moderate quantities of the thyroprotein in the form of iodinated casein (7.5 g/day). This will alleviate at least 50% of the depression in milk yield caused by heat stress but is accompanied by a reduction in bodyweight. Larger doses of thyroprotein are undesirable as they exacerbate hyperthermia beyond the animals' ability to counteract the stress. Further research is needed on the post-thyroprotein administration depression of milk yield that is normally experienced, the long-term effects on milk yield and weight change and individual animal variation in response (Smith and Mathewman, 1986). There is also some evidence (Fourie and Louw, 1968; Terblanche, 1967) that goitrogens will increase feed conversion efficiency in heat-stressed beef cattle, by reducing voluntary feed intake while maintaining growth rate. Kamal, Kotby and El-Fouly (1972) observed a reduced respiration rate with the administration of the goitrogen methyl-thiouracil. Some restoration of milk yield in heat-stressed cattle is also achieved with growth hormone administration (Mohamed and Johnson, 1985).

Nutritional techniques

Encouraging cattle to feed at night rather than during the day can reduce the heat increment to the animal and hence heat stress. This may be done by restricting feed access to the night period and, although this results in reduced intake in the short term, milk production is greater than when feed access is given for 24 h (Richards, 1985). A less extreme alternative is to give the cattle feed access for 24 h but to give fresh feed at night, which will encourage greater intakes during the night compared with the day (Ansell, 1976).

It is important in situations of heat stress to minimize the heat increment of digestion. This can be achieved by feeding energy-rich feeds such as fats that are suitably protected from rumen breakdown and molasses that contains soluble sugars not requiring fermentation. Protein degradability in the rumen should also not be too high or the energetic reactions for detoxification and recycling of surplus nitrogen will add to the heat load of the animal (Chandler, 1987). Heat stressed cattle have a low rumen turnover rate with a low level of total short chain volatile fatty acid production and increased ratio of acetic to propionic acids (Mirtz et al., 1971) which increases protein degradability.

Because of the large water flux associated with the heat stressed condition, restoration of sodium, potassium and bicarbonate levels is necessary. Complex interactions between anions are still being evaluated, but suggest that supplementation with sodium bicarbonate and 1.5% dietary potassium, but not potassium bicarbonate, will be beneficial (Schneider et al., 1984; Mallone et al., 1985).

Photoperiodic manipulation of growth, lactation and voluntary intake

Growth

With cattle there is circumstantial evidence that long days stimulate growth under commercial conditions. In a summary of regular weighing of feed-lot cattle, the Meat and Livestock Commission (MLC) in the UK found liveweight gains to be consistently higher in summer than in winter over a period of 5 years, with intermediate rates during spring and autumn (MLC, 1974). It is unlikely that changes in environmental temperature were responsible for this cycle because the critical temperature of yarded cattle fed *ad libitum* on rations of high digestibility is likely to be lower than winter temperatures in the British lowlands. Nutritional changes were also unlikely to have been directly responsible for the seasonal changes in growth because data were only used from farms which practised *ad libitum* feeding of cereal-based diets. Photoperiod changes in a regular annual cycle and was therefore thought to be the factor which could be responsible for the observed cycles of growth.

Subsequently, properly-conducted experiments have been carried out with sheep, deer and cattle. The most comprehensive work with cattle has been performed by Tucker and his colleagues at Michigan State University. In one experiment heifers exposed to 16 h of light/day (16L (light):8D (dark)) gained weight significantly faster than those exposed to 9–12 hL/day (natural winter days) (Peters *et al.*, 1978). In a further experiment with heifers (Peters *et al.*, 1980), natural winter lighting, 16L:8D and 24L:0D were used. Weight gains were significantly elevated by 16L:8D compared with natural days (0.98 *versus* 0.84 kg/day; $P<0.02$), but not by 24L:0D (0.88 kg/day). Unfortunately, there is no report of carcass weights in these experiments, but heart girth in one of the trials with heifers was increased by 16L:8D, suggesting that the effect was not solely on gut fill.

Weight gain, especially protein deposition, was also stimulated in cattle around puberty (Petitclerc *et al.*, 1983a; Petitclerc, Chapin and Tucker, 1984; Zinn, Chapin and Tucker, 1986). In older cattle short days increased fat deposition (Zinn *et al.*, 1986).

In view of the Michigan results and the MLC survey, further experiments on the practical applications of extended days on the growth and carcass characteristics of cattle should be carried out. As with sheep, continuous lighting does not seem to give optimum growth (Peters *et al.*, 1980); a few farms in the MLC survey left lights on all night and this did not improve weight gains in comparison with those where lights were only used for inspection during the night. It is apparent, however, that positive effects of long daylength have been observed only when lighting is solely by artificial means, or when the natural light is extended with comparatively bright artificial lighting. It is possible, therefore, that a clear contrast between day and night is required and that intensities of 100 lux or less are not sufficient to be registered by animals as being 'day', in contrast to diffuse daylight of a 100 lux or more. Further experiments should, therefore, be carried out on the effects of extending natural days with bright artificial light.

Some work of this type has already been performed by Roche and Boland (1980) who compared the growth and carcasses of cattle in natural light with no extra light or with low (50 lux) or high (250 lux) intensity light from 1600 to 2400 h. They found no effect on silage intake, liveweight gain or carcass weight. In another experiment carried out under conditions of controlled environmental temperature (15°C)

15-month-old cattle were exposed to 16L:8D or 8L:16D and offered silage *ad libitum*: liveweight gain was not affected by treatment, nor was there any difference from a third group exposed to natural (winter) daylengths and temperatures. Two further experiments were carried out with calves by Roche and Boland (1980); there was no effect on weight gains or concentrate intakes, either of 16L:8D *versus* 8L:16D or with natural days, with or without extra lighting from 1600 to 2400 h.

Thus we have conflicting evidence as to whether the extension of short days will stimulate the growth of cattle. There has been more work carried out with sheep and much of this, but not all, supports the contention that growth is stimulated by long days, or by skeleton long days (Forbes, 1982).

Lactation

When cows produce more milk in the summer it is usually ascribed to better nutrition. However, recent work in the USA suggests that long daylength stimulates milk yield (Tucker, 1986). If such a method of stimulating milk secretion is shown to be effective under commercial conditions in the UK it would have advantages over methods involving exogenous hormones or pharmacological agents.

Controlled studies of effects of photoperiod on milk yield have been carried out by Tucker and his colleagues at Michigan State University. Peters *et al.* (1978) found that cows exposed to 16L:8D from calving for 100 days gave 3.1 kg of milk/day more than control animals in natural winter days, subsequent to the peak of lactation. Peters *et al.* (1978) also showed that, between September and March, cows in late lactation gave 8% more milk under 16L:8D than did those in natural daylengths. Further work, reported in more detail (Peters *et al.*, 1981), confirmed these observations with 6.7% higher milk yields and 6.1% higher voluntary food intakes in cows under 16L:8D compared with controls in natural daylengths which varied between 9 and 12 h. It may be significant that the intensity of artificial lighting used in the experiments of Peters *et al.* (1978, 1981) was brighter than that used in most other experiments.

In the only relevant report from British studies, Phillips and Schofield (1989) reported a 3.5 litres/day increase in milk yield with no significant effect on composition; 12 cows in winter days which were extended to 18 h ate more of the complete feed than 12 controls (17.6 *versus* 16.2 kg dry matter (DM)/day). The behaviour of the cows was modified by the light treatment, with a significant increase in the time spent lying, reduced standing and walking times and with greater difficulty in detecting oestrus under extended days. Whether this difficulty was simply due to the cows being more contented or whether there was changed secretion of reproductive hormones was not investigated. The latter is a distinct possibility as the extended light cows tended to have more aggressive interactions than control animals and therefore had plenty of opportunity to express their oestrous behaviour had the potential been present.

Food intake and digestion

Feeding behaviour is influenced by photoperiod. Our observations with dairy cows fed silage in a system in which the feeding behaviour of individuals in a group can be monitored (Forbes *et al.*, 1987) show reduced feed intake at night, the length of the nocturnal period without feeding being positively related to the length of the night (Jackson, Johnson and Forbes, unpublished data).

There are also effects of photoperiod on digestion. Forbes *et al.* (1979, 1981) found increased gut fill in lambs which had been under 16L:8D compared to those under 8L:16D. It has recently been shown that motility of the dorsal rumen and rate of passage of digesta to be faster during the night than in the daytime (Sissons, Gill and Thiago, 1986). If the nocturnal acceleration is related to darkness then long nights would cause reduced gut fill for any given level and pattern of feeding, which would explain the observations on lambs, and have implications for digestion and metabolism, animals exposed to long days having greater digestive capacity and possibly higher digestion coefficients for certain types of feed.

Mechanisms

Comparatively little work has been carried out on cattle which helps to explain the way in which photoperiod affects growth and lactation. However, it is reasonable to assume that the underlying principles are the same as those in laboratory animals and, especially, in the sheep.

Photic information is transmitted by the retina to the hypothalamus, particularly the suprachiasmatic nucleus, thence to the pineal gland which secretes melatonin during darkness. Melatonin then influences the hypothalamus and thereby the anterior pituitary gland. With both lambs and growing cattle, exposure to long days stimulates faster growth than short days (Forbes, 1982) and causes greatly increased prolactin secretion and sometimes reduced plasma levels of cortisol, the latter suggesting reduced stress. A light-induced rhythm in plasma β-endorphin levels has recently been shown by Ebling and Lincoln (1987), long days greatly suppressing plasma levels and increasing pituitary content, again suggesting that long days might reduce stress.

Prolactin has a vital role at the onset of lactation, in the transition of the mammary gland from a growing to a secreting organ, but its importance once lactation has been fully established is less certain in the cow. There is a very marked circennial rhythm of prolactin in ruminant animals and it is clearly established that long days (or a skeletal long photoperiod involving a 'flash' of light in the middle of the night) are responsible for stimulating prolactin secretion. It is important to monitor plasma prolactin in order to ascertain whether a particular photoperiodic treatment has had any biological effect, even though the effects of the increased prolactin levels on the animal may be unclear.

The reduction in cortisol and β-endorphin levels seen in some long-day treatments sugggests, somewhat surprisingly, that this is less stressful than short days, but is consistent with the observation of Phillips and Schofield (1989), that cows spent more time lying when in long days. Thus, there are implications for effects of photoperiod on the welfare of cows and potential difficulties of oestrus detection, both of which need further evaluation.

Following the encouraging results obtained under closely controlled experimental conditions (Forbes *et al.*, 1979, 1981), the extension of short, winter days with artificial light has been investigated as a method of stimulating growth in store lambs, without success (Jones *et al.*, 1982). It was suggested that the relatively low level of artificial lighting was seen as 'moonlight' by the lambs rather than 'daylight'. Increasing the intensity of the lights would be expensive and would not be guaranteed to be effective. An alternative is to expose animals to a 'flash' of light in the middle of the night, which has been shown to increase plasma prolactin in rams (Ravault and Ortavant, 1977). This was subsequently shown to be related to increased growth, whether feeding was restricted (Brinklow, Jones and Forbes, 1984) or *ad libitum* (Schanbacher and Crouse, 1981); these effects could be prevented by removal of the pineal gland (Brinklow and Forbes, 1984). Melatonin levels, elevated during the first part of the night, declined during the flash but did not return to normal nocturnal levels during the second part of the night (Brinklow, Forbes and Rodway, 1984), thus giving a 'long-day' pattern of melatonin.

Melatonin is elevated during darkness in calves also and prolactin concentration in plasma is higher in long days than in short (Stanisiewski *et al.*, 1988b). However, pinealectomy, melatonin infusion or melatonin feeding, all of which affected plasma melatonin levels, had no effect on prolactin and these authors suggested that these two hormones are not causally related in cattle (Stanisiewski *et al.*, 1988b). Prolactin was not elevated by continuous light, as compared with 8L days (Stanisiewski *et al.*, 1988b). Pinealectomy does not completely abolish melatonin secretion in cattle (Petitclerc *et al.*, 1983b; Stanisiewski *et al.*, 1988a).

Once-daily feeding of melatonin to extend the duration of high melatonin levels caused an increase in fat to lean ratio in heifers in one experiment, but not in another in which the cattle were already fat at the start of the experiment (Zinn *et al.*, 1988).

Commercial possibilities

From the above it appears as if extending winter days with artificial light will stimulate growth and milk secretion in many cases. What are the differences between these situations and ones in which no effect has been observed? If it is necessary to maintain a high light intensity then light supplementation might not be commercially attractive.

A nocturnal 'flash' of light appears to be an alternative to extension of daylength. The saving in electricity is not great, however, when compared with the cost of installing lights and time-switches (C. J. C. Phillips, personal communication).

Melatonin can be administered orally or by implant but this would, if anything, depress production. Anti-pineal or anti-melatonin agents should be investigated to aid our understanding of the mechanisms of photoperiod effects but are not likely to be licensed for farm use.

Conclusions

Heat stress is a vitally important limit to animal production in many parts of the world. Although we have many methods of combatting it, the most effective ones are expensive because they require capital investment and/or power and water and therefore difficult to apply in many developing countries. It is unfortunate that ruminants have a digestive system which, although it enables them to utilize poor forages and thus compete less for food with human needs, does produce more heat than the digestive system of non-ruminants.

Manipulation of animal production by photoperiodic means is of particular interest in developed countries and poses a challenge to research workers to unravel the mechanisms involved, but resolution of this puzzle will not have the same impact world-wide as would a solution to the problem of heat stress.

References

Ansell, R. H. (1976) Maintaining European dairy cattle in the Near East. *World Animal Review,* **20**, 1–7

Bianca, W. (1959) The effect of clipping the coat on various reactions of calves to heat. *Journal of Agricultural Science, Cambridge,* **52**, 380–383

Brinklow, B. R. and Forbes, J. M. (1984) The effect of pinealectomy on the plasma concentrations of prolactin, cortisol and testosterone in sheep in short and skeleton long photoperiods. *Journal of Endocrinology,* **100**, 287–294

Brinklow, B. R., Forbes, J. M. and Rodway, R. G. (1984) Melatonin in plasma of sheep subjected to short and skeleton long photoperiods. *Experientia,* **40**, 758–760

Brinklow, B. R., Jones, R. and Forbes, J. M. (1984) The effect of daylength on the growth of lambs. 5. Skeletal long photoperiod. *Animal Production,* **38**, 455–461

Chandler, P. J. (1987) Problems of heat stress in cattle examined. *Feedstuffs,* June 22, pp. 15–16, 35

Ebling, F. J. P. and Lincoln, G. A. (1987) β-Endorphin secretion in rams related to season and photoperiod. *Endocrinology,* **120**, 809–818

Forbes, J. M. (1982) Effects of lighting pattern on growth, lactation and food intake of sheep, cattle and deer. *Livestock Production Science,* **9**, 361–374

Forbes, J. M., Brown, W. B., Al-Banna, A. G. M. and Jones, R. (1981) The effect of daylength on the growth of lambs. 3. Level of feeding, age of lamb and speed of gut-fill response. *Animal Production,* **32**, 23–28

Forbes, J. M., El-Shahat, A. A., Jones, R., Duncan, J. G. S. *et al.* (1979) The effect of daylength on the growth of lambs. 1. Comparison of sex, level of feeding, shearing and breed of sire. *Animal Production,* **29**, 33–42

Forbes, J. M., Jackson, D. A., Johnson, C. L., Stockill, P. *et al.* (1987) A method for the automatic monitoring of food intake and feeding behaviour of individual cows kept in a group. *Research and Development in Agriculture,* **3**, 175–180

Fourie, P. C. and Louw, G. N. (1968) A thyroid depressant as an aid in finishing of feeder steers. *Animal Breeding Abstracts,* **36**, 195

Igono, M. O., Johnson, H. D., Stevens, B. J. and Shanklin, M. D. (1985) Shade and shade plus spray and fan effects on diurnal rectal temperature pattern of lactating Holstein cows during summer. In *Proceedings of the 17th Conference on Agriculture and Forest Meteorology,* May 21–24. 1985. American Meteorological Society, Boston

Jones, R., Forbes, J. M., Slade, C. F. R. and Appleton, M. (1982) The effect of daylength on the growth of lambs. 4. Daylength extension to 20 hrs under practical conditions. *Animal Production,* **35**, 9–14

Kamal, T. H., Kotby, E. A. and El-Fouly, H. A. (1972) Total body solids gain and thyroid activity as influenced by goitrogen diuretics, sprinkling and air cooling in heat-stressed water buffaloes and Friesians. In *Proceedings of the International Atomic Energy Agency, 1972,* Vienna, pp. 177–182

Mallone, P. G., Beede, D. K., Collier, R. J. and Wilcox, C. J. (1985) Production and physiological responses of dairy cows to varying dietary potassium during heat stress. *Journal of Dairy Science,* **68**, 1479–1487

Meat and Livestock Commission (MLC) (1974) *Handbook no. 4. Beef production: dairy-bred calves using cereals and arable products.* MLC, Milton Keynes

Mirtz, F. A., Mishra, M., Campbell, J. R., Daniels, L. B. *et al.* (1971) Relation of ambient temperature and time post-feeding on ruminal, arterial and venous volatile fatty acids, and lactic acids in Holstein steers. *Journal of Dairy Science,* **54**, 520–525

Mohamed, M. E. and Johnson, H. D. (1985) Effect of growth hormones on milk yield and related physiological functions of Holstein cows exposed to heat stress. *Journal of Dairy Science,* **68**, 1123–1133

Peters, R. R., Chapin, L. T., Leining, K. B. and Tucker, H. A. (1978) Supplemental lighting stimulates growth and lactation in cattle. *Science,* **199**, 911–912

Peters, R. R., Chapin, L. T., Emery, R. S. and Tucker, H. A. (1980) Growth and hormonal response of heifers to various photoperiods. *Journal of Animal Science,* **51**, 1148–1153

Peters, R. R., Chapin, L. T., Emery, R. S. and Tucker, H. A. (1981) Milk yield, feed intake, prolactin, growth hormones and glucocorticoid response of cows to supplemental light. *Journal of Dairy Science*, **64**, 1671–1678

Petitclerc, D., Chapin, L. T. and Tucker, H. A. (1984) Carcass composition and mammary development response to photoperiod and plane of nutrition in Holstein heifers. *Journal of Animal Science*, **58**, 913

Petitclerc, D., Chapin, L. T., Emery, R. S. and Tucker, H. A. (1983a) Body growth, growth hormone, prolactin and puberty response to photoperiod and plane of nutrition in Holstein heifers. *Journal of Animal Science*, **57**, 892

Petitclerc, D., Peters, R. R., Chapin, L. T., Oxender, W. D. *et al.* (1983b) Effect of blinding and pinealectomy on photoperiod and seasonal variations in secretion of prolactin in cattle. *Proceedings of the Society for Experimental Biology and Medicine*, **174**, 205–211

Phillips, C. J. C. and Schofield, S. A. (1989) The effect of light supplementation on the production and behaviour of dairy cows. *Animal Production*, **48**, 293–303

Ravault, J. P. and Ortavant, R. (1977) Light control of prolactin in sheep. Evidence for a photo-inducible phase during diurnal rhythm. *Annals of Biology and Animal Biochemistry and Biophysics*, **17**, 1–16

Richards, J. L. (1985) Milk production of Friesian cows subjected to high daytime temperatures when allowed food either *ad lib* or at night-time only. *Tropical Animal Health and Production*, **17**, 141–152

Roche, J. F. and Boland, M. P. (1980) Effect of extended photoperiod in winter on growth rate of Friesian male cattle. *Irish Journal of Agricultural Research*, **19**, 85–90

Schanbacher, B. D. and Crouse, J. D. (1981) Photoperiodic regulation of growth. A photosensitive phase during the light dark cycle. *American Journal of Physiology*, **241**, E1–E5

Schneider, P. L., Beede, D. K., Wilcox, C. J. and Collier, R. J. (1984) Influence of dietary sodium and potassium bicarbonate and total potassium on heat-stressed lactating dairy cows. *Journal of Dairy Science*, **67**, 2546–2553

Sissons, J. W., Gill, M. and Thiago, L. R. S. (1986) Effect of conservation method and frequency of feeding on forestomach motility. *Proceedings of the Nutrition Society*, **45**, 96A

Smith, A. J. and Mathewman, R. W. (1986) Aspects of the physiology and metabolism of dairy cows kept at high ambient temperatures: dissertation review 1. *Tropical Animal Health and Production*, **18**, 248–253

Stanisiewski, E. P., Ames, N. K., Chapin, L. T., Blaze, C. A. *et al.* (1988a) Effect of pinealectomy on prolactin, testosterone and luteinizing hormone concentration in plasma of bull calves exposed to 8 or 16 hours of light per day. *Journal of Animal Science*, **66**, 464–469

Stanisiewski, E. P., Chapin, L. T., Ames, N. K., Zinn, S. A. *et al.* (1988b) Melatonin and prolactin concentrations in blood of cattle exposed to 8 to 16 hours of light per day. *Journal of Animal Science*, **66**, 727–734

Terblanche, H. J. J. (1967) Methyl-thio-uracil in fattening rations of young and mature steers. *Animal Breeding Abstract*, **36**, 34

Tucker, H. A. (1986) Photoperiod influences on milk production in dairy cows. In *Recent Advances in Animal Nutrition*, (ed. W. Haresign and D. J. A. Cole), Butterworths, London, pp. 211–221

Wiersma, F., Armstrong, D. V., Welchert, W. T. and Lough, O. T. (1984) Housing systems for dairy production under warm weather conditions. *World Animal Review*, **50**, 16–23

Yeates, N. T., Edey, T. N. and Hill, M. K. (1975) *Animal Science Reproduction, Climate, Meat, Wool.* Pergamon Press, Oxford

Zinn, S. A., Purchas, R. W., Chapin, L. T., Petitclerc, D. *et al.* (1986) Effects of photoperiod on growth, carcass composition, prolactin, growth hormone and cortisol in prepubertal and postpubertal Holstein heifers. *Journal of Animal Science*, **63**, 1804

Zinn, S. A., Chapin, L. T. and Tucker, H. A. (1986) Response of bodyweight and clearance and secretion rates of growth hormone to photoperiod in Holstein heifers. *Journal of Animal Science*, **62**, 1273

Zinn, S. A., Chapin, L. T., Enright, W. J., Schroeder, A. L. *et al.* (1988) Growth, carcass composition and plasma melatonin in postpubertal beef heifers fed melatonin. *Journal of Animal Science*, **66**, 21–27

Chapter 10

New techniques in cattle housing

M. Kelly and G. B. Scott

Many recent developments in beef and dairy housing have successfully been based on an 'animal orientated' approach to design. This was not always the case. In the 1960s and 1970s increasing pressures on production invariably led to larger herds, managed by fewer people, relying on technology to do the bulk of the work. Systems were inevitably pushed to, and sometimes beyond, their limits in order to increase output. The needs of the stock were not always given sufficient priority.

An 'animal orientated' approach is essential in designing successful housing systems and components of these systems. Such an approach asks a great deal of building designers, since the complex requirements of animals are often poorly defined. Social and personal space for example, are not readily identifiable for cattle. Also deficiencies in the key interactive areas of feeding and resting are likely to lead to social unrest and ultimately poor performance.

A number of building elements are discussed in this chapter, including cubicles for dairy cows, feed barrier design and slatted floors for beef cattle. The importance of observing the behavioural needs of stock cannot be overstated as a prelude to livestock housing design. A good example of this is the 'lunging-space' concept for cubicle design, allowing a cow headspace when it moves forward to rise on its hind legs.

Housing dairy cows

Loose housing in cubicles is the predominant system for in-wintering dairy cows in the UK. Average herd size has more than trebled since the introduction of cubicles, while the labour force has decreased dramatically. On large dairy units, there is a tendency to have demarcation of the work associated with milking, feeding and dung handling and so an overall view and appreciation of the working of the unit may be lost. Cows inevitably do not receive the same individual attention associated with the dairy byre or cowshed.

Since the introduction of quotas, farmers have become even more aware of the need to control production costs in order to maintain profitability. Management aids which minimize labour inputs, such as mechanical scrapers or automatic feed regulation and delivery equipment, are commonplace. However, these must never be used as a means of increasing cow numbers per man, to the extent that good stockmanship and the well-being of cows becomes adversely affected. Putting the design information (as it exists) together as a solution to satisfy the needs of an individual farm is an onerous task. It can only be done with a thorough appreciation of the total farming enterprise. The most appropriate compromise of the many diverse elements involved is the very essence of good building design.

A building decision has a degree of permanence, which is difficult to retrieve should a mistake be made. Modifications are expensive. Errors can be costly and may lead to years of inconvenience when working with stock in the buildings. Good design evolves from the experience of others and involves the many disciplines associated with stock and buildings. Management factors, such as the ability to see and isolate animals, are of paramount importance. The handling of animals should involve veterinary practitioners at the design stage. Slurry handling and storage necessitate an engineering input to ensure that the system works. Yet, amid the multitude of decisions which have to be made, the needs of the animals must come first. The design must be 'animal orientated'.

Loose housing in cubicles offers good opportunities for cows to manifest social interactions and to seek resting and feeding places. Standard layout solutions of the type described by Kelly (1983a) and ADAS (1983) have evolved over the years. The significance of the organization of space however is still not fully understood, e.g. with respect to personal space, refuge space or 'flight distance' (Cermak, 1987a). Deficiencies in design areas of feeding and resting are likely to create social unrest. Ultimately this may lead to poor cow performance.

Cubicle design

The design of the resting area, which is essentially the cubicle, is very important to the well-being of the cow. Inadequate bed length and hard beds are the main reasons why some cows refuse to use cubicles. Recent studies (ADAS, 1983) have related cubicle size to cow weight, chest girth and diagonal body length (Table 10.1). The cow's bodyweight should be based on the average bodyweights of the largest cows, representing approximately 20% of the herd.

Table 10.1 Cubicle dimensions to suit a range of cow bodyweights

Cow bodyweight (kg)	375	425	475	525	575	625	675	725	775	825
Clear cubicle length (m)	2.00	2.04	2.08	2.12	2.16	2.20	2.24	2.28	2.30	2.33
Clear cubicle width (m)	1.10	1.10	1.10	1.20	1.20	1.20	1.20	1.20	1.20	1.20

Courtesy of ADAS.

There is now a greater appreciation of the forward space demand for rising movement and some cubicle divisions cater for this better than others. The 'Dutch Comfort' type (Figure 10.1) allows a cow good headspace when it moves forward to rise on its hind legs. The absence of any side restriction in the pelvic region also means less injury risk when the cow is lying down or rising. Numerous other cubicle division designs satisfy these design criteria, provided that the most important factors of adequate bed length and width are met.

Cows prefer soft, resilient surfaces to bare concrete or to concrete covered in a hard rubber matting. They demonstrate their preference by increased cubicle occupancy time (Tasker and Kelly, 1980; Daelemans, Maton and Lambrecht,

Figure 10.1 Cows lying in Dutch Comfort cubicles

1983). However, once concrete is well-bedded with straw or sawdust, it is much more difficult to establish a preference for the soft carpeting type of product. A well-bedded cubicle is difficult to achieve in practice, because the bedding is continually dragged out of the cubicle. The advantage of a soft carpet is that it ensures a comfortable bed at all times. Use of such products should be encouraged, since the health benefits of having all cows lying in clean, dry, cubicles are considerable.

Feed barrier design

Another zone of great social interaction is the feed area. Performance specifications for feed barriers have been outlined by Burnett and Robertson (1985) and Cermak (1987b). Recommended feed frontages for cattle are given in the relevant British Standard (BSI, 1981).

The most important factor in feed barrier design is to obtain the initial dimensions correctly, so that animals are not competing aggressively for feed. Once this is achieved, then a number of barriers can meet the requirements, including tombstones, diagonals or simple horizontal rails. The barrier must be designed from the starting point of the animal, and not simply for engineering or constructional convenience.

Poorly designed barriers can cause injuries, particularly to the nape and brisket. They can also lead to excessive feed wastage. Typical forces exerted on feeding barriers by a 550 kg dairy cow, were measured at the Farm Buildings Development

Centre in Reading (Cermak, 1987b). It was established that a 20° slope of the barrier allowed cows the maximum forward horizontal reach, while exerting a relatively small force on the barrier itself. The investigation has resulted in a recommended configuration and dimensions for a tombstone feeding barrier for British Friesian dairy cows.

Feeding silage, hay and straw in big-bale form is now commonplace. Robertson, Bain and Burnett (1986) have investigated existing feeders and found 10% wastage of silage in conventional ring feeders, through animals pulling out the silage and trampling it underfoot. A 10% feed wastage level for 100 finishing cattle over a 200 day winter amounts to a total silage loss of approximately 100 tonnes.

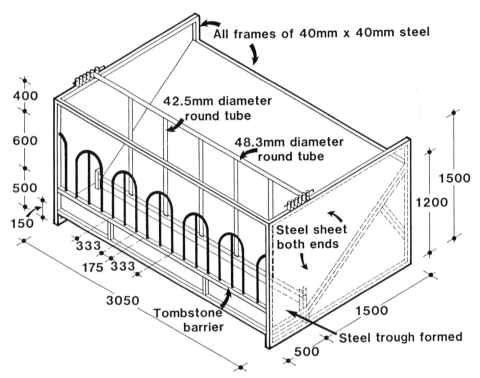

Figure 10.2 A big bale feeder designed to eliminate feed wastage (measurements in mm). (Courtesy of Centre for Rural Building, Aberdeen)

An improved feeder, to hold two big round bales, is now commercially available. The feeder holds two big bales, which are held on a sloping base and slide forward when eaten (Figure 10.2). Any feed dropped by the cattle falls into a secondary trough, adjacent to the tombstone barrier. It cannot be trampled underfoot and so feed wastage is eliminated. The feeder shown is suitable for cattle in the weight range 450–600 kg, kept in groups of 20. The feeder can replace perimeter walling and obviates the need to enter the pen, which is a requirement of the ring feeder. An 'animal orientated' approach was taken by the design team and this worthwhile and practical solution evolved only after a long series of observations on the feeding behaviour of cattle.

Floor design

The area of constant interaction between livestock and the building is the floor. Poorly designed floors can contribute to lameness, but because lameness is multifactorial, it is impossible to assess accurately the importance of the building environment as a cause of the condition. Factors that may be involved include increased trauma, wetness, slippery or abrasive floors, poor foot hygiene, feeding of conserved forage and lack of exercise.

Lameness in dairy cow buildings has been reviewed by a number of researchers, including Kelly (1983b) and Cermak (1987c), who give recommendations for texturing new floors or retexturing slippery and polished old concrete floors. Sensible overall layout design can also help to alleviate the problem. Examples are avoiding sudden changes in levels, eliminating dark and narrow passages and eliminating sharp and potentially dangerous objects like protruding water troughs or gate hinges. Uneven or sloping surfaces greatly increase the risk of falling because cows often display defensive reflexes when confronted with such a situation and make sudden uncoordinated movements (Baggott, 1982).

One design feature which is worth more widespread consideration is the provision of a bedded area in addition to the main cubicle complex. This has proved to be successful when used to cater for the small number of cows not able to cope with the hard floors and standings associated with the cubicle housing system. This is yet again an example of thoughtful 'animal orientated' design, which can succeed, when coupled with good, attentive management.

Housing beef cattle

Many systems for housing beef cattle exist in the UK. These have been summarized by Harper and Johnstone (1983) and by ADAS (1985). Various factors influence the choice of housing system, but one predominant constraint is the availability or otherwise of bedding. Kelly (1985) has demonstrated that there is insufficient straw available to service the housing requirements of the grassland areas in Scotland. Hence the development of minimum-straw or slurry-based housing systems in these areas.

Over the UK the distribution of cattle on grassland does not coincide with the production of straw on arable land. Hence, straw often has to be bought in for bedding and handling costs may be prohibitive. An economic comparison of five housing systems in the UK by Kelly, Smith and Ashworth (1982) is given in Table 10.2.

Table 10.2 Relative gross capital costs and annual cash outflow for five housing systems in the UK

Housing system	Relative gross capital cost	Relative annual cash outflow (straw purchased)
High-level slatted building	100	82
Low-level slatted building	96	82
Slatted building with underslat scrapers to midden	80	71
Bedded court	79	100
Sloped floor, tractor-scraped to midden	73	65

If straw must be purchased then annual cash outflow is affected significantly to the extent that the bedded court becomes the most expensive system to service. The assumption made was that all straw had to be bought in at £30/tonne (1982) and capital repayments were at 15% interest rate. Straw-based systems therefore are not a viable option in the UK if all straw must be purchased. Minimum straw usage is not acceptable within bedded courts since this invariably leads to wet, dirty, lying areas and poor housing conditions, detrimental to the health of the cattle housed.

Sloped floor housing

The main attractions of this system of housing are that it is simple and relatively cheap (*see* Table 10.2). It offers a low or zero straw usage system to those farmers not willing to commit themselves to the high straw demands of a traditional bedded court. The system is still at a relatively early stage of development in the UK, despite being introduced in Orkney about 10 years ago (Robinson, 1984). Parallel developments have taken place in other countries, notably America (Moore and Larson, 1980; Collins, 1982), France (Eoche-Duval, 1982), and Ireland (Flynn and Kavanagh, 1981).

The Orkney sloped floor, described by Robinson (1984) originated in Orkney, but the concept is now known and used throughout the UK. The basic principle involves confining cattle in pens in which the floor slopes away from the feeding trough (Figure 10.3). The slope encourages urine to run off, leaving the higher part relatively clean and dry. Animal activity at the feed barrier has a cleaning effect, in providing a brushing action as animals drag their feet away from the barrier. Stock lie in a double row; parallel and adjacent to the feed barrier.

In small units, all animals are removed from their pens prior to the whole floor being scraped. This can be time consuming for larger units having a number of pens and dividing gates. Since the higher section of the floor, at least 1.5 m from the feed barrier, remains relatively clean, the idea of using a yoked feed barrier has evolved, mechanically scraping the back of the pen. Court width must then be a minimum of 3.8 m to allow for tractor access.

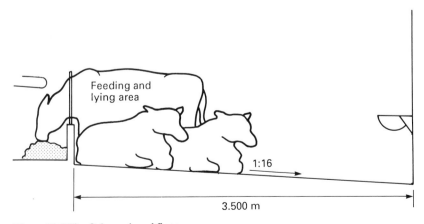

Figure 10.3 The Orkney sloped floor

Even with this modification, there is still the necessity for daily scraping. This can be overcome by the installation of an automatic scraper (Figure 10.4). The scraped area contains the dung and helps to define the drier lying area. It may be possible to reduce the dung passage width to give an overall court width of 3.5 m, running scrapers more often to cope with the increasing depth of dung.

Figure 10.4 Sloped floor with an automatic scraper, showing straw laid adjacent to the feed barrier

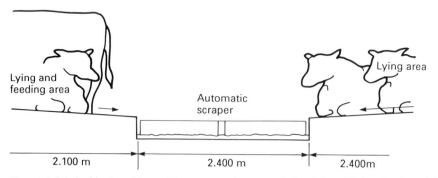

Figure 10.5 A double sloped floor with an automatic scraper designed for *ad libitum* feeding of silage

A further development involving automatic scrapers is to allow one scraper to service two slopes (Figure 10.5). This suits an *ad-libitum* feeding system and is successfully in operation in one unit in Dumfriesshire. Competition for feed does not affect the lying behaviour of cattle at the barrier, since cattle have the option to lie undisturbed on the slope at the back of the court.

Figure 10.6 Reversed sloped floor with a lying area at the back of the pen and a scraped area adjacent to the feed barrier. Note the slotted roof which has greatly improved the ventilation of this building

A further variation comprises a scraped feed stance with a stepped lying area at the back of the court (Figure 10.6). This system, with an automatic scraper, has been incorporated into the Netherwood steading, Crichton Royal Farm, West of Scotland College. Unlike the Orkney sloped floor, the reverse sloped floor does not rely on the brushing action of feet at the barrier to clean down the slope, but merely on the activity associated with the standing and lying on that area. Indications from the Netherwood steading are that the slope is self-cleansing, provided that stocking density is right and bedding is minimal. The scraped feed stance is automatically scraped twice a day, and this assists greatly in keeping cattle clean.

Experience of sloped floors indicates that injuries to stock are rare and no udder problems have been encountered with suckler cows. Feed type has an important bearing on the success of the system, since it affects both animal cleanliness and dung handling. High dry matter feeds tend to result in cleaner cattle, but frequency of scraping is also important in this respect. Twice a day scraping is recommended. Stock can have faeces adhering to their coats, particularly in the early part of the housing period. However, animals do perform well, despite not always looking clean.

Slatted housing

The relatively high capital outlay of slats has affected their use in new buildings in recent years, although in the UK if a farm is in a less favoured area, the slats and tanking can now receive a 60% grant. The quality of the slats themselves has improved considerably, due both to better manufacturing techniques and more realistic design criteria, including a provision for lateral as well as vertical loading (BSI, 1987).

Foot and leg injuries associated with cattle movement on slats continue to cause concern. Good management helps by batching cattle of similar sizes in small groups, with good feed and water access. Nevertheless, it is a harsh, uncompromising housing system, because of the hardness of the floor and injury problems that can occur as slats become worn and slippery.

In an attempt to improve cattle comfort on slats, Irps (1983, 1987) has investigated the use of softer floor coverings. This work has included the use of rubber-coated slats and trials have been carried out in pens floored by rubber-coated slats or a combination of rubber-coated and conventional slats. Results are not conclusive. Irps claims that the behaviour of cattle on rubber-coated slats indicates they are more comfortable, but lack of abrasion results in more hoof growth, which can itself cause difficulties.

Rubber-coated slats for housing dairy calves are highly recommended by Suss (1987). Daily weight gain on the softer slats was higher than on the conventional slats. Heat loss from the animals to the floor was reduced on the rubber-coated slats. On the rubber-coated slats conditions favoured claw overgrowth, while on conventional slats, hoof wear rate exceeded growth rate. If unchecked this would lead to injury. However, a major defect in the rubber-coated slats was noted by Suss, in that although the manufacturers claim that the lifetime of the rubber-coated slats exceeds 6 years, damage to the slats was observed after only 2 years.

Another approach to cattle comfort on slats has been taken by Boxburger and Pfadler (1980a,b). Present stocking densities of approximately $2\,m^2$/animal are considered too high, since injury risk is increased and performance adversely affected, but too low a stocking density can lead to dirtier cattle. A wide slot will improve cleanliness but can lead to an increased risk of claw injury, since for 80% of footsteps part of the foot is on a slot. Therefore a compromise arrangement is to have narrow slots *and* slats with low stocking densities. Stained leather, resulting from dirty hides, costs the German leather industry several millions of Deutschmarks each year, so it is important to have finishing cattle as clean as possible.

Cattle cleanliness on slats and in other housing systems is currently being investigated by Scott and Kelly (1989). Cleanliness itself is not considered in the UK Welfare Code for Cattle (MAFF, DAFS and WOAD, 1983), and it is misleading directly to relate lack of cleanliness to poor welfare. Clean animals however, look healthier and do not incur financial penalties at the market.

To overcome the subjectivity of cleanliness, a scoring system was devised, as shown in Figure 10.7 with a typical cleanliness score.

A clean area of the coat is given a score of zero while an extremely dirty area can score up to three. This simple technique was shown to be repeatable and scores ranged widely, even for the same system of housing (Scott and Kelly, 1989). No one floor type is exclusively cleaner than any other. Poorly bedded courts, for example, can lead to very dirty animals, while animals on slats can be very clean, or moderately dirty, depending on many factors. These factors include cattle size, density of stocking, coat length, feed type, management, and frequency of bedding and scraping.

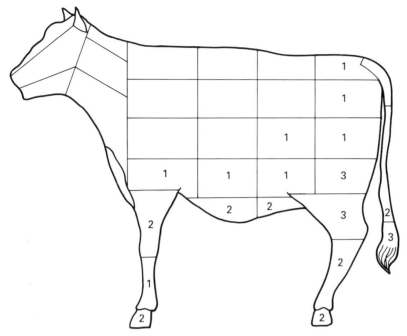

Figure 10.7 A cleanliness scoring system for beef cattle showing a typical score during winter housing

Scott and Kelly (1989) are currently investigating a radical design of a slatted floor, which attempts to give maximum comfort, without unacceptably compromising cleanliness. The conventional slat width has been increased to minimize the number of gaps and hence provide a surer footing for cattle, with less risk of unsupported hoof injury. In addition, using the self-cleansing principle of the sloped floor, the slat surface has been sloped to a 1:16 gradient. The slats are 600 mm wide and are abutted alternately to form a bed width suitable for a lying animal (Figure 10.8). To improve further the comfort aspect of the floor, a rubber surface has been added, comprising of grooved rubber matting. However, before the trial slatted pen was totally covered in rubber matting, half the pen was covered for a small trial to investigate any animal preference for surface finish.

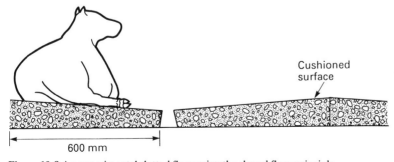

Cushioned surface

600 mm

Figure 10.8 An experimental slatted floor using the sloped floor principle

Observations suggest that the rubber-coated slats were used for lying in preference to the concrete ones. Lying position is influenced to some extent by the geometry of the pen, in that animals do tend to lie in the quiet area of the pen, away from feed barrier activity. Work is continuing by attempting to increase comfort by surfacing the rubber with Enkamat.

The ventilation of cattle housing

The promotion of good natural ventilation has been a consistent feature of livestock building design work for a number of years. A variety of techniques has been developed to encourage building manufacturers and designers to provide a healthy environment for stock. One example is the protected open-ridge design, initially developed by Bruce (1978) and subsequently refined as illustrated in Figure 10.9. This design enables an outlet opening to be retained, without the problem of rain ingress, usually onto a central feed pass. Outlet and inlet design can then be balanced to encourage the 'stack effect' ventilation, essential to good airflow, even on still days.

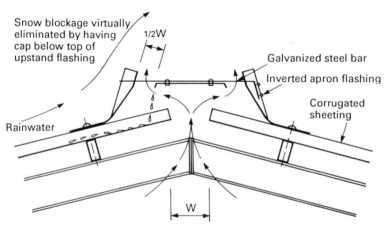

Figure 10.9 A protected open ridge design based on work done at the Centre for Rural Building, Aberdeen

One further technique, which is proving to be very successful, is the spaced or slotted roof. Design methods for this and for raised roof sheets have been published by Bruce (1977, 1978, 1981). It is particularly applicable to wide span, multispan buildings or buildings in close proximity to others, through which good air distribution can often be a problem. The technique involves the slotting or spacing of roofsheets to create openings approximately 20 mm wide down the length of the roof (see Figure 10.6). Hence a 'breathing' roof is formed, which acts as both inlet and outlet. The openings are formed in the crown of the sheets, so creating a channel either side to run off rainwater. A bright, healthy, internal environment is created which is of benefit to both stock and management. It is a technique which is very much to be encouraged in promoting the good natural ventilation of stock buildings.

A recent technique developed for extension purposes by the Scottish Agricultural College Building Service is a computer-based assessment of ventilation requirements for cattle buildings. Information is given on the openings required at different stocking densities (Centre for Rural Building, 1987), which enables farmers to modify a wide range of buildings to promote a healthy environment for the livestock.

Conclusions

Refinements to existing systems and elements of cattle housing are continuing and these are improving animal well-being and comfort. The loose housing of dairy cows in cubicles will continue as the dominant housing system. A greater understanding of behavioural needs of cows has led to improved cubicle and feed barrier designs in recent years. 'Animal-orientated' work of this type must be encouraged if we are to make design improvements.

Beef cattle housing systems are tending towards simple, low-cost solutions. This has led to some interest in the sloped floor concept, especially in areas short of straw. Many variations are possible, but good attentive management is essential in keeping animals clean. The sloped floor concept has been used by Scott and Kelly (1988) to investigate a new type of cattle slat. So far, results are inconclusive, although a rubber coating on the floor seems to promote greater animal comfort. This is backed up by recent work in Germany, especially by Suss (1987) observing young cattle on rubber-coated slats.

Good natural ventilation of cattle buildings is continually being promoted and computer-aided design now assists greatly in quickly determining outlet and inlet openings for a wide range of buildings. The slotted or spaced roof is a well proven concept, which provides a 'breathing roof', ensuring an excellent airflow distribution.

Putting the range of building design information together as a built solution to satisfy the requirements of an individual farm is an onerous task. The ability to compromise the many, and sometimes conflicting, elements involved is the very essence of good design. Attentive stock management is paramount to the success of any enterprise. The job of the building designer is to facilitate attentive stock management by consistently taking an 'animal-orientated' approach towards the design solution, using the best information available. The job of those involved in building research and development is to provide that information.

References

Agricultural Development and Advisory Service (ADAS) (1983) *Design and Management of Cubicles for Dairy Cows*, Booklet 2432. 30 pp
Agricultural Development and Advisory Service (ADAS) (1985) *Beef Cattle Housing*, Booklet 2512. 61 pp
Baggot, D. G. (1982) *Farm Buildings Association Technical Journal*, **30**, 12–15
Boxberger, J. V. and Pfadler, W. (1980a) Untersuchungen zur Ausfuhrung von Spaltenboden fur Milchkuhe. (Investigation into the design of slatted floors for dairy cows). *Landtechnik*, **5**, 227–231
Boxberger, J. V. and Pfadler, W. (1980b) Anforderungen an Spaltenboden in Liegeboxenlaufstallen. (Requirements for slatted floors in cubicle stalls). *Landtechnik*, **1**, 41–44
British Standards Institution (BSI) (1981) BS 5502: Section 2.2

British Standards Institution (BSI) (1987) BS 5502: Part 22

Bruce, J. M. (1977) *Slotted Roofs for Cattle Buildings. A Design Aid.* Scottish Farm Buildings Investigation Unit Leaflet

Bruce, J. M. (1978) Protected open-ridge design. *Farm Building Progress*, **53**, 9–10

Bruce, J. M. (1981) Design method for natural ventilation of cattle buildings using raised roof sheets. *Farm Building Progress*, **66**, 17–21

Burnett, G. A. and Robertson, A. M. (1985) Cattle feeder design – a methodology. *Farm Building Progress*, **82**, 21–25

Centre for Rural Building (1987) *Ventilation of Cattle and Sheep Buildings*, Ventilation Design Leaflet, Centre for Rural Building, Aberdeen

Cermak, J. (1987a) Design and management of confinement housing for dairy cows. Its effects on behaviour and incidence of lameness and injuries. *Paper presented to the American Association of Bovine Practitioners.* Seminar, Phoenix, 26 pp

Cermak, J. (1987b) The design of feeding barriers and mangers and its effect on incidence of injuries and feed wastage. *Paper presented to the American Association of Bovine Practitioners.* Seminar, Phoenix, 8 pp

Cermak, J. (1987c) Design of slip-resistant surfaces for dairy cattle buildings. *Paper presented to the American Association of Bovine Practitioners.* Seminar, Phoenix, 8 pp

Collins, W. H. (1982) Young stock housing. In *Proceedings of the First Annual Farm Builders Conference and Tour*, Pennsylvania

Daelemans, J., Maton, A. and Lambrecht, J. (1983) *Farm Buildings Digest*, **18**, 10–12

Eoche-Duval, J. (1982) Inclined floors with straw littering. *La France Agricole* 1937, 13–20

Flynn, V. and Kavanagh, A. J. (1981) Comparison of slatted floor and solid concrete housing systems for beef cattle. *Preliminary Report.* Agricultural Institute, Grange

Harper, A. D. and Johnstone, K. (1983) Building options for housing beef cattle. *Farm Building Progress*, **74**, 17–27

Irps, H. (1983) Results of research projects into flooring preferences of cattle. In *Farm Animal Housing and Welfare*, (ed. S. H. Baxter, M. R. Baxter and J. A. D. MacCormack), Martinus Nijhoff, The Hague, pp. 200–215

Irps, H. (1987) The influence of the floor and keeping system on the behaviour, the claw condition and weight gain of fattening bulls. In *Latest Developments in Livestock Housing*, CIGR Section 2 Seminar, University of Illinois, pp. 22–26

Kelly, M. (1983a) Cow cubicle layout within the dairy complex. *Farm Building Digest*, **18**, 9–15

Kelly, M. (1983b) Good dairy housing – a form of preventive medicine. *Veterinary Record*, **113**, 582–586

Kelly, M. (1985) Current development in beef cattle housing in Scotland. *Farm Buildings and Engineering*, **3**, 29–32

Kelly, M., Smith, P. G. and Ashworth, S. W. (1982) An economic comparison of some cattle housing systems. *Farm Building Progress*, **70**, 13–36

Ministry of Agriculture, Fisheries and Food, (MAFF), Department of Agriculture and Fisheries for Scotland, (DAFS), and Welsh Office Agriculture Department, (WOAD) (1983) *Codes of Recommendations for the Welfare of Cattle*, Leaflet 701, 16 pp

Moore, J. A. and Larson, R. E. (1980) Manure movement and cattle preference on different solid sloping floors. *Transactions of the American Society of Agricultural Engineers*, 964–967

Robertson, A. M., Bain, C. W. and Burnett, G. A. (1986) Big bale feeder for cattle. *Farm Building Progress*, **84**, 9–12

Robinson, T. W. (1984) The Orkney sloped floor. *Farm Building Progress*, **78**, 11–14

Scott, G. B. and Kelly, M. (1989) Cattle cleanliness in different housing systems. *Farm Building Progress*, **95**, 21–24

Suss, M. V. (1987) Kalberaufzucht auf gummiummantelten Betonspaltenboden. (Rearing calves on a rubber-coated concrete floor). *Landtechnik*, **4**, 153–154

Tasker, R. J. and Kelly, M. (1980) Cow cubicle bed treatments: Salviacim and Enkamat K. *Farm Building Progress*, **62**, 3–6

Chapter 11

New techniques in cattle health control

D. W. B. Sainsbury

It is especially appropriate when considering new techniques in cattle production to use the positive word 'health' rather than the more negative one 'disease'. For the future our thinking, planning and actions must be predominantly concerned with the promotion of good health in cattle. Good health is the birthright of every animal and we now have better means at our disposal to achieve this than ever. Certainly much has been achieved in recent years in either eliminating totally or greatly reducing the consequences of several devastating contagious and infectious diseases. Conditions such as foot-and-mouth disease, tuberculosis and brucellosis have been all but eliminated from the British Isles; nevertheless this is not the position world-wide.

It is estimated (Blajan, 1985) that the world-wide financial costs of foot-and-mouth disease are $50 000 million, brucellosis $3500 million and mastitis, which is still very much with us, costs $35 000 million (Table 11.1). Also, as fast as

Table 11.1 Estimates of world-wide costs of leading animal diseases ($ million)

Foot-and-mouth	50 000
Rift valley fever	7 500
Shipping fever	3 000
Neonatal diarrhoea	1 750
Blue tongue	3 000
Bovine enzootic leucosis	900
Leptospirosis	4 500
Brucellosis	3 500
Mastitis	35 000

From Blajan (1985).

some of the more virulent easily diagnosable and relatively simply eliminated diseases are dealt with, there is a tendency for their replacement by conditions which are difficult to diagnose, slow to develop and often subclinical so that their effect may not be recognized at all.

Virtually all farm livestock perform well below their full genetic capabilities due to the harmful effect of pathogens whose presence can very well go unrecognized. It is the objective of good recording schemes and improved diagnostic techniques to alert the farmer to what is happening at the very earliest stages. By the use of this information he may well be able to apply the proper corrective action for the shortest possible time and at the minimum cost. There are likely also to be important lessons guiding the farmer towards preventive measures perhaps by the use of medicines, but more probably by biological means, such as the use of vaccines, and above all by carrying out appropriate adjustments in management. However, the fundamental measures that should always be used to ensure good health are often ignored. This may be simply because of ignorance; alternatively it may be that they are treated with contempt even when they are known. A further, all too common reason is that there is a growing belief that modern technology in disease control allows the farmer to dispense with the fundamentals altogether and he can call on all the sophisticated armoury of drugs, antibiotics and vaccines to maintain good health in his stock. Nothing could be further from the truth. Although the fundamentals of health control go back for as long as stock have been kept intensively, there is new knowledge and techniques associated with them which very rightly come within the field of this chapter.

Hygiene

When considering hygiene there is a lesson to be learnt from the great successes achieved in the UK in eliminating the major animal plagues – controlling or eradicating them even before the cause of the disease was known. The techniques used were based on acquired knowledge of the epidemiology of the disease which was sufficiently accurate to achieve its eradication even in ignorance of its aetiology (Henderson, 1983). The dreadful cattle scourge of rinderpest was finally eradicated in 1877 by the application of principles elaborated more than 150 years before by Thomas Bates, surgeon of George I, and a Commission of Justices of the Peace of the County of Middlesex (Animal Health, 1965). These principles are worth repeating even now: any suspicion of illness had to be reported; sick cattle had to be slaughtered and buried; cow byres had to be washed and left empty for two months; persons attending sick animals were prohibited to go near healthy stock and the sale of sick cattle was prohibited.

All these measures are still applicable to the present day but we can now be far more efficient. It was the beginning of sound practice to wash a building and leave it empty for some time, during which a great many of the disease causing agents would die out without the presence of the normal living host. There is no need for us to leave buildings empty now because we have detailed knowledge of the range of pathogens, or potential pathogens, which are likely to be present and we know which disinfecting agents can destroy them. If we apply this information properly there is no need to leave a building empty except to wash and dry it. If the pathogens have not been destroyed with the disinfectant, they may persist for months or years either protected by organic matter or, as is very often the case, living within the bodies of vectors ranging from insects to rodents.

Depopulation of buildings and pasture

A further principle that was adopted before it was fully understood why it was so effective was the 'all-in, all-out' system. It has become essential in the most successful rearing of intensively kept livestock, used almost without exception with growing poultry and pigs, but increasingly recognized as essential in the successful rearing of calves and now increasingly adopted for housed animals of all ages. The same principles apply to the outdoor animal even though it obviously has to be applied in a rather different way.

The policy of 'all-in, all-out' with a rigid disinfection procedure between batches of animals is much more important in young animals because they are often highly susceptible to infection while their immunological mechanisms develop. Under practical conditions we can never remove the challenge from pathogens altogether though this may remain an ultimate aim. With calves, the benefits of the 'all-in-all-out' system are greatly appreciated and these may be appreciated sufficiently for it to become essential in all cattle rearing carried out by intensive methods.

Site size

If the benefits of an 'all-in, all-out' procedure are accepted then there are several other practical implications of quite profound importance. The ideal may be to empty the whole site of stock after every batch. If this can be practised it usually means that the size of the site will be limited. Those who have carried this out are usually well satisfied with the consequences, but every step towards the policy, even without following it absolutely, is likely to be worthwhile. If the site cannot be emptied, then the buildings may be given a complete rest. If the buildings cannot be emptied completely it may be possible to subdivide them by solid partitions and deal with each part in turn. All such steps reduce the challenge and lower the risks.

Thus it is apparent that from the point of view of livestock health and well-being, the smaller the better. Of course, there has to be a balance between obtaining optimal biological results and the achievement of economic efficiency. Britton and Hill (1975) in their careful study of size and efficiency in farming concluded that farms of quite modest size achieve the highest efficiency and the conclusion may be drawn that a farm with two or three stockmen is as good as any from most aspects and especially that of animal health. They also emphasized that there is little evidence that any appreciable economies of size are generally to be gained by enlarging beyond the point where the farmer remains in very close contact with his employees, if indeed he has any at all outside the family.

It should be the aim to have a livestock unit which is in harmony with its immediate environment and one essential of great importance from the health aspect is that all the effluent can be safely disposed of so that it is not a hazard to health and can be put to good use on the land. This one criteria alone will tend to favour units of modest size.

Vaccines and medicines

At the turn of the century there were great strides made in promoting the health of cattle by the development of effective vaccines and they could often form an invaluable precursor or adjunct to a specific disease eradication programme which would be completed with a slaughter policy. However, when chemotherapeutic and antibiotic medicines became available later in this century, the development and use of vaccines took a back-seat and the veterinary profession was beguiled into attempting to cure rather than prevent disease. It may be considered this was a great mistake, for while these new medicines enabled great strides to be made in the control of animal diseases, they could never by themselves be effective in eliminating them. They did, however, persuade the livestock industry that disease could now be controlled and so livestock enterprises became much more intensive. Rather stupidly it was never really understood that a healthy animal would always be at an economic advantage to an animal with pathogens apparently contained by medicines. Very often, in any event, the diseases were no more than rendered subclinical by using medicines, so that depression of productivity could still be extremely serious.

There followed the Swann Report (1969) which concluded that a much more careful use of medicines was required and, from that time, restrictions have been put on the use of therapeutic medicines, generally supported by the pharmaceutical industry itself (AVI Symposium, 1981). That is not to suggest that medicines do not have an important part to play in promoting the good health of cattle but no system of cattle management should ever be dependent on drugs. If it is, it may be predicted that it will fail.

It is also important that we are fully aware of the consumers' concern about residues of drugs in the food, backed up by some demanding legislation (Crawford, 1988). Though a certain amount of the reaction against the use of medicines is exaggerated and emotional and the dangers have rarely been established, there is no doubt about the emergence of drug-resistant pathogenic organisms. The animal industry should step well clear of being criticized on these grounds. A good example of the progress achieved with medicines is afforded by considering mastitis in the dairy cow.

Annual mastitis cell counts in England and Wales decreased from just under 600 000/ml in 1971 to 352 000/ml in 1986, but in 1988 the trend has increased so that it is 14 000 cells/ml higher than 12 months ago. It is to be hoped that this is not a sign of a serious increase in mastitis (Booth, 1988a,b), though it does suggest that in dairy management, as in that of other livestock, too many corners are being cut. In this respect it is pertinent to draw attention to the adverse effect on the environment created by bad cubicle design, poor cleaning and litter management, ineffective drainage of passages and harsh, especially damp, conditions in the housing and collection and dispersal areas. Many of the new techniques developed in the housing field have improved the efficiency of handling the beasts and materials, but have sadly been at the expense of the good care of the livestock.

The new technology

Livestock health has as yet been little influenced by the emerging new biotechnologies, but they are beginning to be applied and can be expected to do so increasingly in the future. These new techniques will involve the prevention, treatment and diagnosis of disease. Recombinant DNA technology is the source of the majority of the developments and includes:

(1) the production of vaccines, hormones and therapeutic proteins
(2) DNA probes used as diagnostic tools
(3) the production of amino acids, vitamins and antibodies.

New vaccines based on this technology have, for example, been developed against the foot-and-mouth disease virus and colibacillosis diarrhoea in calves. The benefits of the new vaccines are that they are much safer, produce a better and uniform immunity and create a minimum of stress on the animal. In the area of diagnostics, DNA probes will enable a more rapid detection of pathogens. This will be particularly useful in clinical tests for conditions which have a very complex background such as respiratory disease of growing cattle or mastitis in the cow. They also promise to be more finely specific in their action and the test can more readily distinguish a vaccine from a disease challenge.

Therapeutic proteins in the form of interferons and lymphokines may offer a useful tool in the treatment of infectious disease and especially complex virus infections.

In the area of diagnostics, the production of monoclonal antibodies will provide very finely tuned techniques for the diagnosis of diseases, expressed as 'finger-printing' the specific agents causing the disease. Such methods will be of the greatest help in achieving an early diagnosis of conditions that we are attempting to eradicate and possibly in the detection at the earliest stage of worrying diseases of unknown aetiology such as bovine spongiform encephalopathy.

Genetic control

Scientific investigations have been carried out for many years on the prospects of controlling disease by selection for genetic resistance. To date these have not been very successful, partly because the use of other methods of disease control are generally considered quicker and cheaper and partly because the commercial advantages of concentrating on selecting for improved production have outweighed those of selecting for disease control. This is likely to continue to be the position in the short term. In the long term, however, genetic selection for disease control is likely to emerge as a measure of great importance. It will be much easier to achieve than hitherto because the diagnostic tools made available by the new technology will make it much less complicated, aided by the use of computerized information on health and productivity. The tantalizing possibilities have already been considered on the control of mastitis by genetic means (Solbu, Spooner and Lie, 1982). The case for the breeding of disease-resistant stock has been admirably summarized by Payne (1983). Genetic measures will, however, be much more important in selecting cattle that will be less susceptible to the metabolic stresses and strains that are inevitable as productivity continues to improve.

Conclusions

We appear to be on the threshold of an exciting new era in the control of infectious diseases of livestock even though the majority of the developments are still in their early stages. However, there is not the slightest indication that the new biotechnology will replace the need for absolute attention to the fundamentals of good management. Indeed, to make use of the new techniques the rules of good management will need to be rigorously updated. Throughout the world pharmaceutical and related companies are pouring enormous investments into the development and production of new diagnostic biological and therapeutic products. If the livestock industry is going to be ready to harvest the benefits there is an urgent need for local veterinary services to prepare for the inevitable changes. It is suggested that there will be a need for a relatively small number of veterinary teams, perhaps regional and mobile, staffed with highly qualified professionals and with the range of support equipment required. This will be much more fitting than the larger number of low-technology laboratories currently favoured. Government funding for veterinary research is being quite savagely cut in many developed countries and the only way to overcome the disadvantage this may bring is to develop a suitable degree of cooperation between research centres, universities and commercial interests, combined with the further education of the practising veterinary surgeon which is already underway. The 'sharp end' of health control will inevitably remain in the hands of the local veterinary practice, but it must have the backing of highly efficient laboratories.

Just one further feature would be highly desirable and that is proper national and international assessment of disease prevalence. A modern disease survey should ensure that efforts are concentrated in the appropriate areas and there would be correlation drawn up of disease incidence with its effect on the economics of production. In the UK with all the benefits that modern technology can offer and taking advantage of the enormous benefits that our island isolation ensures, we should be able to make the elimination of most of the common infectious cattle diseases a reasonable ambition. Internationally this is a much more long-term goal, but still feasible in the way that human infectious diseases are being brought under control.

References

Animal Health: a Centenary 1865–1965 (1965) HMSO, London

AVI Symposium (1981) *Ten Years on From Swann*. Proceedings of Symposium organized by the Association of Veterinarians in Industry, 1981. Unwins, London

Blajan, L. (1985) Production and utilization of food and fibre products. *Proceedings XXIII World Veterinary Congress*, pp. 4–46

Booth, J. M. (1988a) Update on mastitis. I. Control measures in England and Wales. How they have influenced incidence and aetiology. *British Veterinary Journal*, **144**, 316–322

Booth, J. M. (1988b) Disappointing mastitis trend. *Veterinary Record*, **123**, 167

Britton, D. K. and Hill, B. (1975) *Size and efficiency in Farming*. Saxon House, Farnborough

Crawford, I. (1988) Residues – drugs and growth promoters. *Journal of the Royal Society of Health*, **4**, 225–226

Henderson, W. H. (1983) Priorities for research. In *The Control of Infectious Diseases in Farm Animals*. British Veterinary Association Trust Project, pp. 50–55

Payne, L. N. (1983) Breeding resistant stock. In *The Control of Infectious Diseases in Farm Animals*. British Veterinary Association Trust Project, pp. 37–42

Solbu, H., Spooner, R. L. and Lie, O. (1982) *Proceedings of the 2nd World Congress on Genetics Applied to Livestock Production*, Madrid, **VII**, 368–382

Swann Report – Joint Committee on the use of Antibiotics in Animal Husbandry and Veterinary Medicine (1969) Cmnd 4190, HMSO, London

Chapter 12

Recent developments in the provision for cattle welfare

J. L. Albright and W. R. Stricklin

New herd management developments have important effects on the well-being of cattle, and the cattle enterprise is very well suited for the application of electronics and automation (Albright, 1987). Automation, considered by some to be detrimental to the welfare of animals in intensive units, may not necessarily be so. The time saved, together with reduced drudgery, could free people for more human:animal interactions allowing better care. Many minor faults could be corrected before major breakdowns occur. Automation can increase the time given to the training of stockmen and take the pressure off existing workers which, in turn, might reduce accidents and injury. In addition, animals in large intensive units may be treated more humanely if stress to the worker is reduced (Kilgour, 1985).

Intensification of the dairy industry has not necessarily resulted in reduced cow welfare. For example Israel has the highest average milk production per cow (9250 kg) of any country in the world (Raz, Ezra and Ron, 1987). While the larger Kibbutz herds are being milked in shifts three times daily (often with three different shifts of personnel making milking a less demanding job), family-owned herds (Moshav) are being milked only twice-daily. A third milking increases individual cow production by about 8–15% and also means more frequent handling and observation of the dairy herd, sometimes accompanied by an increased intake of feed, and cows being more frequently cooled off in the milking parlour and holding pen etc.

The inventiveness of Israeli dairymen has come a long way in reducing the stress that accompanies the extreme temperatures and high humidity found throughout Israel. Thus far, economic constraints rather than genetic potential or technical knowledge seem to slow further progress as expressed by individual and herd average yields.

Other recent management developments that have improved the comfort and well-being of cattle include the raising of calves in individual pens or hutches (Baker, 1981); providing exercise for cows prior to calving (Lamb *et al.*, 1979); type and location of resources, e.g. smooth material for feeding areas at floor level (Albright, 1983a), lock-in stanchions (Albright, 1987); plentiful supply of drinking water (Anderson, 1984); new designs in cubicles (Irish and Merrill, 1986; Jarrett, 1986); shade and cooling (misting) during hot weather (Armstrong *et al.*, 1984, 1985; Schultz, Collar and Morrison, 1985); adequate bedding to allow for cow comfort and insulate the udder against extremes in cold, while keeping cows dry, clean, out of the mud and free from flies and pests (Albright, 1983b); feeding at night to influence daytime calving (Pennington and Albright, 1985); and elimination of stray voltage (Lefcourt, Kahl and Akers, 1986; Albright *et al.*, 1988).

For as long as cattle have been milked there has been the art of cow care that results in more milk from healthier, contented cows. For some time now, it has been recognized that the dairy cow's productivity can be adversely affected by discomforts, and the importance of contented cows has been emphasized. Good dairy farmers avoid management practices that place undue stress on their cows because mistreated animals perform poorly. Alert herdsmen have the perception and ability to read 'sign language' in animals. Also, in many cases, cattle placed under duress show signs by bellowing, butting or kicking. These behavioural indications of adjustment to the environment are always useful signs that the environment could be improved. In some cases, the way animals behave is the only clue that stress is present (Stephens, 1980; Albright, 1983a; Curtis, 1983).

Some standard agricultural practices with temporary painful procedures can be justified on the basis of sustaining long-term well-being. Recommendations for beef cattle husbandry include dystocia management, castration, dehorning, and cattle handling procedures. In addition, identification methods, supernumerary teat removal, foot care, milking machine and udder sanitation are recommended for dairy cattle (Curtis *et al.*, 1988).

Developments in reproduction

In many herds poor reproductive performance in cows may be due to being confined continually on concrete surfaces. Ideally, for good oestrous detection, cows should be allowed access to a dirt exercise lot twice-daily for a minimum of an hour each time (Britt, 1982). There is generally an immediate increase in mounting activity as soon as cows get onto dirt. Oestrous detection rate and oestrous duration are higher on dirt lots than concrete (Britt, Armstrong and Scott, 1986). Cows that have slipped on wet concrete continue to be reluctant to mount other cows once they leave concrete.

Grooving concrete reduces injury (Albright, 1981a, 1982) while giving cows confidence during walking and mounting. Much new construction now incorporates slopes of 1% or less since slopes of 2–5% are too severe for walking and mounting. Continuous observation, additional research and field trials are needed to provide further answers to questions on concrete conditions, slope and cow management. Since many people lack the necessary background, attitude, perception and caution when dealing with dairy bulls and parturient cows, additional training plus cow and bull behavioural information are needed (Arave and Albright, 1981). Milk production is a byproduct of the reproductive process, and a multidisciplinary approach has been advocated for research, paying particular attention to 'integrated reproductive management' (McGinnis, 1982).

In an adequate environment heifers are normally mounted between 80 (measured in USA) and 120 (measured in UK) times during an oestrous period of average duration 14h (USA 18.2h and UK 10h). In total approximately 1000 oestrous-related activities are initiated and received during each oestrous cycle, including successful and unsuccessful mounting, chin-resting on the rump, and sniffing and licking the anogenital area. The frequency of sexual activity is reduced by heat stress and low body condition (Pennington, Albright and Callahan, 1986) and during other essential activities such as feeding and milking (Pennington, Albright and Callahan, 1985). In addition Pennington *et al.* (1986) have shown that high yielding cows that exhibit a large increase in milk yield and decrease in bodyweight post-partum have reduced sexual activity levels.

Reproductive efficiency in the widest possible sense, reflects the level of adaptation (Tschantz, 1978; Zeeb and Beilharz, 1980). Selection for reproductive yield under defined rearing conditions should lead to requirements appropriate to the species. However, if the environmental conditions for rearing animals are continually being changed, animals are not able to adapt. Behavioural research results (Donaldson, Albright and Black, 1972) suggest that reproduction is optimum when calves are reared and fed individually in separate isolated pens as compared to competitive rearing and feeding in groups.

Slats for dairy cattle have been criticized in recent years due to foot problems, poor oestrous detection and lack of cow comfort. Cattle walk differently when placed on slats (i.e. 'duck-walk' with their rear legs). There is an absence of assertive behaviour, possibly because animals have much less confidence on slatted floors than on other types of flooring (Lees, 1962; Hill et al., 1973).

Due to lack of data and as a safeguard, many dairymen remove their cows completely from concrete during the dry period. From an animal health standpoint, forced exercise during the dry period improves calving ease and reduces udder oedema and retained placentas, but has no effect on subsequent 305 day lactation yields (Lamb et al., 1979). During lactation cows need only a small amount of exercise since eating, ruminating and producing milk results in considerable work, metabolism and expenditure of energy. Continuing forced exercise after parturition therefore reduces feed intake and milk yield. In good weather, though, many dairy farmers move milking cows out of stalls to lots or pasture areas daily to 'sun' and groom each other and themselves (about an hour each day), to stretch and move and to exercise especially their feet and legs (Albright, 1981b). These cows, in turn, are more likely to show signs of oestrus, health and overall well-being. It is interesting that during sunny or mild winter weather, over 95 years ago like today, dairy farmers turned their cows out into an open yard (Plumb, 1893).

During the dry period, cattle should be provided with clean, dry housing or surroundings. The cows prior to calving will seek a quiet separate place to calve. A pasture that is dry and well drained is an excellent location. Maternity pens should be of ample size, be clean, dry and draft-free. Research at several institutions is in progress to influence the time of parturition through feeding time (Muller et al., 1989). If calves could be born during the daylight hours, assistance could be rendered as necessary at a more opportune time. By varying the time of feeding of a total mixed ration, Pennington and Albright (1985) were able to increase the proportion of dairy cows calving between 05.00 and 21.00 h to 85% from the expected 66%.

Developments in social behaviour

Dairy calves

The use of individual pens indoors or outside calf replacement hutches in dairy heifer raising is well-documented (Baker et al., 1981). In the work of Fisher et al. (1985) housing system did not influence bodyweight gain or feed conversion prior to weaning, but postweaning calves in pens 0.66 m wide with grated floors had lower weight gains (0.74 versus 0.90 kg/day) and less favourable feed conversion (2.00 versus 1.77 kg dry matter intake/kg bodyweight gain) than calves housed in pens 1.36 m wide with solid floors bedded with straw. Hutches or dome-shaped units (Johnson, Reneau and Otterby, 1985) allow for individual care, decrease morbidity and mortality, and eliminate intersucking. During daylight hours on the coldest days, pens with high density polyethylene around them were approximately 11°C (20°F) warmer than the surrounding air and 8°C (15°F) warmer than plywood hutches.

Some researchers (Donaldson, 1970; Donaldson, Albright and Black, 1972; Arave and Warnick, 1979; Arave, Mickelsen and Walters, 1985) have found significant increases in milk production in cows from calves raised in isolation compared with those raised and fed in a group. Additional research is underway to clarify whether this is an effect of imprinting, isolation stress, social isolation or primary socialization (Creel and Albright, 1987). Data collected for 28 calves over their first year included bodyweight, behaviour and urinary cortisol while in the rearing environment, serum cortisol to measure the response to stress as heifers, serum progesterone to determine age at puberty, measures of socialization to humans, dominance relationships, and medical records. The influence of neonatal isolation on production does not appear to be due to effects on dominance, primary socialization, or reproductive maturation.

A fourth hypothesis proposes that social deprivation is stressful for calves and causes chronic elevation of cortisol. Chronic neonatal adrenocorticoid elevation has been shown to increase the potential of the hypothalamo-hypophysial-adrenal system to respond to acute stress. Neonatal isolates might be physiologically better able to deal with environmental stress as adults, resulting in greater milk production. The cortisol concentrations of isolates up to 10 weeks of age tended to be higher (3.2 ng/ml) than controls (2.5 ng/ml). Subsequently, increased adrenocortical reactivity or isolates was demonstrated by subjecting yearling heifers to acute stressors (roundup, placement in test arena, approach scores, haltering and handling, and blood sampling). Isolates had a mean serum cortisol concentration of 13.2 ng/ml compared with 9.4 ng/ml for controls. While in the rearing environment, isolates tended to stand, vocalize and investigate more than controls, which might indicate environmental stress. Isolates were lighter than controls as heifers, a difference approaching statistical significance ($P<0.10$) possibly indicating reduced growth through the influence of chronic stress on production and secretion of growth hormone. Isolates and controls experienced roughly equal numbers of treated illnesses during the study. Immune suppression through corticosteroid elevation was therefore not indicated. These tests and research by others support the hypothesis that isolation stress has an organizational effect on the ontogeny of the hypothalamo-hypophysial-adrenal system of neonatal calves. The resultant stronger response to adult stressors could increase milk production (Creel and Albright, 1987).

Dairy cows

Dairy cows exhibit a strong social hierarchy in the herd. This is usually stable but factors such as crowding can tend to destabilize it. It is unclear what constitutes crowding in dairy cow environments (Baker et al., 1981). Reduction of the cubicle and eating place allowance per cow from 1.0 to 0.8 has been shown to increase aggressive interactions and disrupt the dominance hierarchy (Wierenga, 1984). Metz and Wierenga (1984) also showed that reducing the space allowance for a herd leads to increased aggression and that these effects were strongly dependent on the amount of time the cows were allowed to adjust to the new population density.

To reduce aggressive interaction low-producing dominant cows should be culled. Hog rings can be placed in the poll of obvious dominant cows to eliminate fighting at the feed bunk and all cows should be dehorned as an aid to reducing aggressive interaction. Weak submissive cows should be culled as well as low-producing dominant cows in order to have a more stable herd (McFarlane, 1976).

It is important to shift cows from one group to another only when necessary and preferably to move small groups of cows (Hart, 1980; Arave and Albright, 1981). Not only is there social pressure on a cow in a new group, but she may have different amounts of feed, a new herdsman milking and a different milking time. Group size should be kept stable and no larger than 100 cows (Albright, 1978a).

Cattle to man relationships

A good relationship with cattle is based upon communication as well as confidence. Competent cowmen talk to their cows when the cows are under stress. They use a pleasant voice yet display the necessary confidence. A good communicator with cows is quiet rather than excitable and reinforces good behaviour in his cows by pleasant words and a reassuring touch. A good herdsman does not shout at, frighten or strike his cows. It has been suggested that the best test for cowmanship is whether the cows come to the herdsman in the pasture or turn away and run as he approaches. Cows in the high producing Israeli herds are often literally touching each other and record breaking cows like Beecher Arlinda Ellen, the world's record milk producer from Rochester, Indiana with 25 270 kg of milk in one year, are always very tame (Albright, 1978b, 1981b).

As creatures of habit, gentle dairy animals may be excited into rebellion by the use of unnecessarily severe methods of handling and restraint (King, 1978; Lemenager and Moeller, 1981).

According to Grandin (1985), rough handling of farm animals is not only inhumane, but can cause excessive losses due to sickness and slower growth. Careful husbandry and handling of livestock in all phases of production are a prerequisite to a profitable business (Curtis et al., 1988).

Calves that have been reared by their dams or by nurse cows with no human involvement are more difficult to calm down, have greater flight distances (15 m), circle continuously in the holding pen, and are difficult to train to the milking routine. This approach seems to be a poor way to save on labour (Albright, 1982).

Developments in resting behaviour

Dairy cow comfort has been a concern of dairymen for a very long while. Most types of housing situations will work provided that the cows are given adequate space and the system is well-managed. The dairy cow has a relatively long life compared with other major classes of livestock and it is an objective of dairymen to increase as much as possible the productive life of the dairy cow. No small part of the environment required for the promotion of longevity is the comfort of the cow (Larsen, 1982). There is a tendency to reduce her space in order to reduce costs. Space reduction should not be allowed to reduce cow comfort. Taking into account her comfort, studies have revealed that a cow in a 126 cm × 215 cm 'comfort' stall (cubicle) produces 1 kg more milk per day than when housed in a smaller stall, 108 cm × 169 cm. Comparing these 'comfort' stalls with regular tie stalls showed that cows in 'comfort' stalls remain cleaner, spend almost two hours more per day lying down and also have fewer knee, hock and udder injuries. Stalls that are too short can also give rise to feet and leg problems (Longhouse and Porterfield, 1954).

Cows aim to achieve a rather fixed amount of lying, and their well-being must seriously be impaired when lying time is restricted for several hours (Metz, 1985). When Holstein cows studied over a 4-week period were given either $9\,m^2$ or $2.25\,m^2$ per cow or housed in an 183 cm × 231 cm isolation pen they produced equal quantities of milk (Arave, Albright and Sinclair, 1974). Milk leucocyte levels, recorded to measure stress (Whittlestone *et al.*, 1970), tended to increase during isolation. Surprisingly, the average leucocytes found were higher for cows at $9\,m^2$ than for cows at $2.25\,m^2$. Corticoids, another measure of stress were not changed appreciably (Table 12.1).

Table 12.1 Milk production, milk leucocytes, plasma corticoids and behavioural comparisons of cows in normal and restricted space

	9 m²/cow	2.25 m²/cow
Milk production (kg/day)	16.0	15.7
Milk leucocytes (no/ml)	626 000	463 000*
Blood plasma corticoids (ng/ml)	7.0	4.0
Herdmate encounters (fights)	279	189*
Yard zones entered (2.4 m squares)	788	341*

* Differences were significant ($P < 0.01$).

Arave, Albright and Sinclair, 1974.

Over a 2-year period Hill *et al.* (1973) compared cows housed in cubicles on a clay base with a sawdust–shavings bedding (and concrete passageway) with cows on a slatted passageway and indoor–outdoor carpeting placed on concrete-filled cubicles. Those with sawdust–shavings bedding produced about 1.4 kg more milk per day. They had lower leucocyte counts in 15 of the 20 months, stayed cleaner, had less clinical mastitis and fewer feet and leg injuries. They were more comfortable as they spent about an hour more lying down during the night. They also spent less time standing half in the cubicle and half in the passageway which is indicative of less comfort.

Also over a 2-year period, Crowl (1952) observed differential responses to the type of bedding used in a stanchion barn. Employing the number of times observed standing as an index of discomfort and observing cows at 15-min intervals for a total of 34 times during the night, he found that cows were least comfortable when ground or crushed corncobs were used and most comfortable when chopped wheat straw was used. Wood shavings and sawdust gave intermediate responses. Hacker *et al.* (1969) used the same technique as Crowl (1952) and found marked behavioural preferences in terms of cow usage for cubicle mats. Lactating cows preferred synthetic resin-over-rubber cow mats for 79% of the time spent lying. Foreman *et al.* (1958) showed that variability in the behaviour of individual cows in comfort tests appeared to be more influential on resting habits than did type of stall (comfort, tie or stanchion).

Behavioural responses of cows to straw yard and cubicle housing under the same roof in a complete confinement system were compared by Schmisseur *et al.* (1966). Cows in a straw yard exhibited more group action while animals in cubicles were more individualistic. About 40% of cows showed a significant preference for a specific cubicle while 60% used the cubicles at random.

Cows prefer to stand or lie uphill and a slope of 5–8 cm for a cubicle is quite adequate. Neck boards or rails (107 cm or more from the floor), which encourage cows to step backward as an aid to keep the bedding free from contamination, are in routine use in cubicles. One major problem is when they are set too low (<77 cm) from the floor as the cow is crowded with less 'head room'. The Mid West Planning Service (Bates *et al.*, 1985) suggests 91 cm from the floor for suspended partitions while others recommend 107–122 cm (Irish and Merrill, 1986; Jarrett, 1986). There are new designs in cubicle divisions to keep the cow's hips from getting stuck, with greater ease of rising and better cow orientation ('Dutch comfort design'). Many dairy farmers want their cubicles to be management-free, but without careful attention cows can develop poor temperaments and vices. Experience reveals that it is best to correct vices early, because once the habit is established it is extremely difficult if not impossible to stop.

Over a 24-h period, a cow may lie down eight to ten times (Cermak, 1977). Normally, after about 2 h spent resting, she will get up and lie down again, often on the other side of her body. About 55% of resting time normally occurs between 2200 and 0400 h.

Passageway flooring should be firm and allow the cow to walk with confidence. It is suggested that cows confined to concrete continuously be given frequent exercise in a dirt lot or pasture (Albright, 1983a). Heifers and cows find preferential areas within the housing environment for mounting and displaying sexual activities (Pennington, Albright and Callahan, 1985, 1986). Not enough is known quantitatively about how hoof growth and wear are affected by different housing and flooring situations. Cubicles should have proper bedding to allow for cow comfort and insulate the udder against extremes in cold. Before making a sizeable investment in some form of rubber tyres (Hodgson *et al.*, 1985), solar Enka mats (Moore and Moore, 1985), poured concrete cubicles, wood platforms, asphalt, rolled rubber belting, vulcanized (compressable) rubber, indoor–outdoor carpeting, polyester (plastic) mat, synthetic resin, metal perfalot, rubber mats, cow pillows, etc, dairy producers should see the material installed and working, preferably under winter conditions.

McGuffey (personal communication, 1980) found that given the choice of earth, bridgeplank, perfalot or concrete surfaced cubicles, cows disliked concrete the most. Natzke, Bray and Everett (1982) gave cows the choice of carpets or a layered mat of polyester, polyvinyl chloride and nylon materials, in cubicles and cows preferred the carpeted cubicles. In a subsequent trial (Natzke *et al.*, 1982) a vulcanized (compressable) rubber product with layered mat was preferred by cows over solid rubber cow mats. Cows produced the most milk when given a choice of cubicles, which also corresponded to the period with the greatest time of stall utilization. In a recent Canadian winter trial (Newberry and Fisher, 1988), cows preferred cubicles with soil for bedding over cubicles with rubber tyres on them (13.1 h compared to 7.4 h per cubicle per day spent lying).

Cows adjust readily to their surroundings and feeding areas, but care must be given to dehorning them humanely as calves and to keeping them dry, clean, free from flies and pests, cool and out of mud (Albright, 1981b, 1982). According to Fraser (1980) stress may arise from systems which make excessively high demands of the animals, i.e. physical restraint or 'boredom'. Ewbank (1978) has speculated that rumination acts as an 'anti-boredom' activity in the adult bovine. Considerable self-stimulation and 'inwardness' occurs in cattle due to the rumination process. During rumination, whether lying or standing, cows are quiet, appearing to be relaxed with their heads down and their eyelids lowered. It is possible for cows to ruminate while standing, but they usually lie down, with their chests against the ground.

During illness or oestrus, rumination time is reduced. Cows spend much less time sleeping than people, dogs, and horses. Cattle are drowsy for about 8 h/day (Ruckebusch, 1972). Whereas most non-ruminants sleep on their sides, it is necessary for the ruminant to maintain an upright posture (sternal recumbency) so that the oesophageal opening to the rumen does not drop below the fluid level (Balch, 1955), which would prevent eructation of rumen gasses and causes bloat. As a result, ruminants only sleep for a few minutes at a time.

Developments in feeding

Domestic cattle respond differently to various types of feeding, and herdsmen can study the behaviour of animals and use this knowledge to increase production. For instance, feeders and watering systems must be placed where the young or inexperienced animal can find them. Management methods based on cost per man hour or per animal rather than the optimal productive efficiency of animals often dictate practices. Accessibility of feed may be more important to animals than the actual amount of nutrients provided. Efforts must be made to reduce the competition for water, minerals and shelter. Also space, density, and distribution of feed are closely associated factors. The following suggestions are aimed at helping the cow in large herds and enabling the herdsman to reduce stress.

Fenceline feeding or auger feeding of complete feeds (total mixed rations) should be practised to allow all cows to feed at once and reduce aggression. Holstein cows fenceline fed a complete ration of silage and concentrates ate for 26% more of the time following feeding than the same size group eating from a trough which they had to travel around (Albright, 1974). Many visitors to the Purdue Dairy Research Center, Indiana facilities, where weighed, complete blended (maize silage plus concentrates) rations were first fed with a mixed wagon over 20 years ago, have commented on how tame, docile and relaxed the cows appear to be (Howard et al., 1968). Uptake of this feeding system has been inhibited by the lack of knowledge about how to formulate complete diets (Owen, 1979).

Many dairies practise fenceline feeding with cows eating with their heads in the natural grazing-like position. During her world milk production record, American Holstein Beecher Arlinda Ellen ate hay at floor level. Thre is evidence that cows eating with their heads in the downward position produce 17% more saliva than cows eating with their heads held horizontally which has a direct influence on the efficiency of rumen function (McFarlane, 1972).

Concrete mangers can be renovated with epoxy-like finishes or relined with wood or tiles to aid feed consumption. Over time, silages with their low pH tend to etch concrete thus exposing the cow's tongue and mouth to rough edges and stones. Deep (46 cm) troughs facilitate once-daily feeding and ease labour requirements on weekends. Conversely shallow (30–46 cm) troughs require more frequent distribution of feed which results in fresher feed. Cows exhibit more rooting behaviour in shallow, elevated troughs. For more than a year, milking cows at

Purdue University were observed about once-weekly, for feed-tossing behaviour. About 10% of the cows participated in this rooting, sorting ritual. Feed wastage from feed tossed over their backs or along their sides was up to 10% each week. Feed-tossing occurred especially in summer during heavy fly concentration but took place in winter as well. When cows were given the alternative of eating from an elevated bunk with the floor and top of the trough 28 and 76 cm from ground level respectively, or from the same trough at ground level, they chose the lower level. The group fed at ground level showed no feed tossing behaviour. It appears that this is a livestock engineering problem remedied easily by feeding in the natural head-down, grazing-like position.

The floor of the trough should be level or with no more than 1% slope along the length of the bunk. With troughs on slopes of 3–5% or more there is a shift and movement of cows in the direction of the slope. The cows keep shifting and moving. Also, with excessive slopes of 5% or more in holding pens, cows eventually become hesitant about entering the area as well as the milking parlour (Albright, 1983a).

When maize silage is the only forage fed, hay should be provided, allowing 2.3–4.5 kg of long hay per cow. Accompanying the move to all or high maize silage diets in the USA has been an increase in digestive upsets, displaced abomasums and fat cow syndrome. Cows fed all maize silage diets or feed chopped too finely (Sniffen and Chase, 1981) do not ruminate for as long as those on hay diets. All maize silage diets need a great deal of buffering plus added protein to compare favourably with alfalfa hay.

After milking, cows should return to fresh feed. This keeps the cow from lying down and helps to finish udder drying. Lying down immediately after milking can result in bacteria entering the streak canal through the open teat sphincter and eventually coliform mastitis. Most top dairy farmers prepare cows' udders before milking and dip their cows with recommended teat dips after milking.

In a study by Anderson (1984) when eating time was restricted to 2.5 h at each feeding with no water during the first 2 h, feed consumption, milk yield and water intake were significantly lower than 24 h feed and water availability. Anderson also found that when pairs of tied-up cows share a water bowl the dominant cow eats, drinks and yields significantly more than the submissive cow. It was hypothesized that the submissive cow may suffer from chronic stress. The only way to be certain in all situations is to have one water bowl per cow.

Feeding in the parlour may induce the cow to enter, but will result in more dust, flies, and defecation. Feeding equipment in the parlour (as well as the parlour itself) should be properly grounded to avoid any stray voltage shocks. The dramatic behavioural responses displayed by cows subjected to electric shock are not correlated with significant or prolonged physiological responses. This dichotomy, although probably exaggerated in cows, suggests that electrical shock may not be a good paradigm of stress (Lefcourt, Kahl and Akers, 1986). Following the installation of a tingle voltage filter on 10 problem Indiana dairy farms there was a 10-fold decrease in stray voltage from 0.7 V to 0.06 V. At this level the cow is not at risk (Albright et al., 1988). Controlled experiments on stray voltage (Appleman, personal communication, 1986) have shown that minimal cow response is expected for neutral-to-earth voltage differentials of less than 0.36 V. Voltage in excess of 1.0 V may significantly alter animal behaviour and reduce milk yields even when shock duration is brief (0.5 s).

Developments in locomotory behaviour

No matter how rough a concrete floor is when laid, constant scraping with a mechanical blade will soon give it a fine polish causing cows to slip. Also, cows in concrete yards and cubicle buildings appear to be 'walking on eggs'. It is a very difficult task manually to roughen up a large area of concrete floor with chisels and hammers. Mechanical grooving with a scaled down version of a highway grooving machine prevents skid accidents, cow loss or injury. Before grooving at Purdue (West Lafayette and DuBois, Indiana) in the autumn of 1978, about 1.5% of dairy cows were lost each year due to slipping on the smooth concrete. Since the grooving was undertaken, no cows have been culled due to debilitating injury (especially after calving) on the slippery concrete (Albright and Hill, 1985). Two cows were injured away from the grooved area where there is a hazardous slope. Cows in oestrus are now mounting one another again in the cubicle alleys. Cows seek out the grooved area and they are more active and show greater confidence in their walking especially to, from and in the collection yard (Albright, Hill and Moeller, 1981).

In situations where concern for the potential for infectious necrobacillosis of the hoof (footrot) outbreaks or other foot problems is significant, the use of footbaths containing 5–10% of either formaldehyde or copper sulphate or 10% copper sulphate in slaked lime is advised (Blood, Radostits and Henderson, 1983). Such footbaths should be placed in areas of heavy traffic flow such as exit areas from the parlour. As these conditions are likely to arise under environmental conditions that are adverse to general health and management of stock, attention should be paid to removing the predisposing causes of foot problems (i.e. moist or muddy environments underfoot).

Future research in cattle welfare

Observation of cattle has been going on for centuries (Albright, 1987) and helps to increase knowledge and improves husbandry techniques. A more logical approach to the study of cow behaviour and training is now advocated linking it with the commercial situation. A knowledge of normal behaviour patterns with perception provides an understanding about cattle and results in improved management that will achieve and maintain high yields. Cattle must fit in well with their herdmates as well as their handlers. Proper mental attitude of the herdsmen must blend in with skillful management and humane care in today's highly competitive, technological society (Albright, 1978b).

References

Albright, J. L. (1974) Let cow sociology help you plan a feeding system. *Successful Farming*, **72** (7) D1–4

Albright, J. L. (1978a) Social considerations in grouping cows. In *Large Dairy Herd Management*, (ed. C. J. Wilcox and H. H. Van Horn). University of Florida Press, Gainesville

Albright, J. L. (1978b) The behaviour and management of high yielding dairy cows. In *The Behaviour and Management of High Yielding Dairy Cows*. BOCM Silcock Dairy Conference. Heathrow, UK. January 30. 44 pp. Booklet

Albright, J. L. (1981a) Dairy cattle housing and management. *Dairy Science Handbook,* **14**, 95–100

Albright, J. L. (1981b) Dairy industry developments that improved the welfare of dairy cows and veal production. *Feedstuffs,* **54**, 23–33, 43

Albright, J. L. (1982) Behavioural responses to management systems – dairy. In *Proceedings Symposium on Management of Food Producing Animals,* (ed. W. R. Woods). Purdue University, West Lafayette, vol. I, pp. 139–165

Albright, J. L. (1983a) Putting together the facility, the worker and the cow. *Proceedings of the Second National Dairy Housing Congress,* American Society of Agricultural Engineers, St Joseph, Michigan, pp. 15–22

Albright, J. L. (1983b) Status of animal welfare awareness of producers and direction of animal welfare research in the future. *Journal of Dairy Science,* **66**, 2208–2220

Albright, J. L. (1987) Dairy animal welfare: current and needed research. *Journal of Dairy Science,* **70**, 2711–2731

Albright, J. L., Dillon, W. M., Sigler, M. R., Wisker, J. E. *et al.* (1988) Dairy farm analysis of stray voltage problems in Indiana. *Journal of Dairy Science,* **71** (suppl. 1), 215 (abstract)

Albright, J. L. and Hill, D. L. (1985) Livestock engineering – the interaction between facilities, management and behaviour. *Midwest Animal Behaviour Society Meeting,* Miami University, Oxford, Ohio, p. 1 (abstract)

Albright, J. L., Hill, D. L. and Moeller, N. J. (1981) Grooving 15-year old concrete improved footing. *Hoard's Dairyman,* **125**, 744

Anderson, M. (1984) Drinking water supply to housed dairy cows. *Sveriges Lantbruksuniversitet Diss. Rpt. 130,* Uppsala, 123 pp

Arave, C. W. and Albright, J. L. (1981) Cattle behaviour. *Journal of Dairy Science,* **64**, 1318–1329

Arave, C. W., Albright, J. L. and Sinclair, C. L. (1974) Behaviour, milk yield and leucocytes of dairy cows in reduced space and isolation. *Journal of Dairy Science,* **57**, 1497–1501

Arave, C. W., Mickelsen, C. H. and Walters, J. L. (1985) Effect of early rearing experience on subsequent behaviour and production of Holstein heifers. *Journal of Dairy Science,* **68**, 923–929

Arave, C. W. and Warnick, V. D. (1979) Heifers reared in isolation milked well at Utah State. *Hoard's Dairyman,* **124**, 618–619

Armstrong, D. V., Wiersma, F., Fuhrmann, T. J., Tappan, J. M. *et al.* (1985) Effect of evaporative cooling under a corral shade on reproduction and milk production in a hot, arid climate. *Journal of Dairy Science,* **68**, (suppl. 1), 167 (abstract)

Armstrong, D. V., Wiersma, F., Gingg, R. G. and Ammon, D. S. (1984) Effect of manger shade on feed intake and milk production in a hot arid climate. *Journal of Dairy Science,* **67**, (suppl. 1), 211 (abstract)

Baker, F. H., Beck, A. M., Binkert, E. F., Blosser, T. H. *et al.* (1981) *Scientific Aspects of the Welfare of Food Animals.* 35 pp. Council for Agriculture Science and Technology, Ames, IA, Report no. 91

Balch, C. C. (1955) Sleep in ruminants. *Nature (London),* **175**, 940–941

Bates, D. W., Bickert, W. G., Brugger, M. F., Bodman, G. R. *et al.* (1985) Milking herd facilities. In *Dairy Housing and Equipment Handbook.* Midwest Plan Service, Iowa State University, Ames, IA, p. 46

Blood, D. C., Radostits, O. M. and Henderson, J. A. (1983) Infectious footrot in cattle. In *Veterinary Medicine.* 6th edn. Balliere Tindall, London, pp. 662–664

Britt, J. H. (1982) Foot problems affect heat detection. *Hoard's Dairyman,* **127**, 824

Britt, J. H., Armstrong, J. D. and Scott, R. G. (1986) Estrous behaviour in ovariectomized Holstein cows treated repeatedly to induce estrus during lactation. *Journal of Dairy Science,* **69** (suppl. 1), 91 (abstract)

Cermak, J. P. (1977) The behaviour of dairy cows. *Farm Buildings Progress,* **49**, 19. Translation from *Ethology of Domestic Animals* by J. Hauptman, Statni Zemedelski Nakladatelstvi, Praha (in Czech)

Creel, S. R. and Albright, J. L. (1987) Early experience. In *The Veterinary Clinics of North America: Food Animal Practice – Farm Animal Behavior,* (ed. E. Price). Saunders, Philadelphia, pp. 251–268

Crowl, B. W. (1952) Bedding cows with different materials. *Hoard's Dairyman,* **98**, 879 and 893

Curtis, S. E. (1983) Environment and animal behaviour. In *Environmental Management in Animal Agriculture.* Iowa State University Press, Ames

Curtis, S. E., Albright, J. L., Craig, J. V., Gonyou, H. W. *et al.* (1988) Guidelines for beef cattle husbandry and guidelines for dairy cattle husbandry. In *Guide for the Care and Use of Agricultural Animals in Agricultural Research and Teaching.* Consortium, Champaign, IL, USA

Donaldson, S. L. (1970) The effects of early feeding and rearing experiences on social, material and milking parlor behaviour in dairy cattle. *PhD dissertation.* Purdue University, West Lafayette

Donaldson, S. L., Albright, J. L. and Black, W. C. (1972) Primary social relationships and cattle behavior. In *Proceedings of the Indiana Academy of Science,* **81,** 345–351

Ewbank, R. (1978) Stereotypies in clinical veterinary practice. In *Proceedings of the First World Congress on Ethology Applied to Zootechnic,* Ministerio du Agricultura, Madrid. 1, 499

Fisher, L. J., Peterson, G. B., Jones, S. E. and Shelford, J. A. (1985) Two housing systems for calves. *Journal of Dairy Science,* **68,** 368–373

Foreman, C. F., Curry, N. H., Homeyer, P. D. and Porter, A. R. (1958) A comparison of different stalls for dairy cattle. *Iowa State College Journal of Science,* **33,** 43–53

Fraser, A. F. (1980) In *Farm Animal Behaviour,* 2nd edn. Bailliere Tindall, London

Grandin, T. (1985) Livestock handling needs improvement. *Animal Nutrition and Health,* **40,** 6

Hacker, R. R., Albright, J. L., Taylor, R. W. and Hill, D. L. (1969) Cow preferences for permanent bedding materials supported by different foundations in a free stall slatted floor barn. *Journal of Dairy Science,* **52,** 918 (abstract)

Hart, B. L. (1980) Bovine behaviour. *Bovine Practice,* **1,** 8–9

Hill, D. L., Moeller, N. J., Hungblut, D. H., Parmelee, C. E. *et al.* (1973) The effect of two different free stall housing systems upon milk production, milk quality, health and behavior of dairy cows. *Journal of Dairy Science,* **56,** 668 (abstract)

Hodgson, A. S., Riley, R. E., Jr, Harrison, J. H. and Murdock, F. R. (1985) Free stall surface materials. *Journal of Dairy Science,* **68** (suppl. 1), 238 (abstract)

Howard, W. T., Albright, J. L., Cunningham, M. D., Harrington, R. B. *et al.* (1968) Least-cost complete rations for dairy cows. *Journal of Dairy Science,* **51,** 595–600

Irish, W. W. and Merrill, W. G. (1986) How we would design free stalls. *Hoard's Dairyman,* **131,** 74

Jarrett, J. (1986) Environmental effects on mastitis and milk quality. *Proceedings of the Dairy Free Stall Housing Symposium,* Northeast Agricultural Engineering Service, Ithaca, NY, p. 31

Johnson, D. G., Reneau, J. K. and Otterby, D. E. (1985) *Minnesota Dairy Report 1985–1986*

Kilgour, R. (1985) The definition, current knowledge and implementation of welfare for farm animals – a personal view. In *Advances in Animal Welfare Science,* (ed. L. Mickley and M. Fox). The Humane Society of the United States, Washington, DC, pp. 31–46

King, J. O. L. (1978) In *Introduction to Animal Husbandry.* Blackwell, Oxford, p. 29

Lamb, R. C., Baker, B. O., Anderson, M. J. and Walters, J. L. (1979) Effects of forced exercise on two-year-old Holstein heifers. *Journal of Dairy Science,* **62,** 1791–1797

Larsen, H. J. (1982) Bedding materials for dairy cattle. In *Proceedings of the Symposium on Management of Food Producing Animals,* (ed. W. Woods). Vol. II, pp. 834–843. Purdue University, West Lafayette

Lees, J. L. (1962) Dairy cows on slats. *Agriculture,* **69,** 226

Lefcourt, A. M., Kahl, S. and Akers, R. M. (1986) Correlation of indices of stress with intensity of electrical shock for cows. *Journal of Dairy Science,* **69,** 833–842

Lemenager, R. P. and Moeller, N. J. (1981) Cattle management techniques. In *Handbook of Livestock Management Techniques,* (ed. R. A. Battaglia and V. B. Mayrose). Burgess Publishing Company, Minneapolis

Longhouse, A. D. and Porterfield, I. D. (1954) Dairy cows need larger stalls. *West Virginia Agricultural Experimental Station Circular no 91*

McFarlane, I. S. (1972) Bovine behavior patterns. *Livestock Breeders Journal,* December, p. 6

McFarlane, I. S. (1976) A practical approach to animal behavior. *Dairy Science Handbook,* **9,** 67–73

McGinnis, W. C. (1982) Federal funding of dairy research in the 1980s. Producer involvement – a must. *Journal of Dairy Science,* **65,** 680–682

Metz, J. H. M. (1985) The reaction of cows to short-term deprivation of lying. *Applied Animal Behaviour Science,* **13,** 301–307

Metz, J. H. M. (1985) and Wierenga, H. K. (1984) Spacial requirements and lying behavior of cows in loose housing systems. *Proceedings of the International Congress on Applied Ethology in Farm Animals,* University of Kiel, Federal Republic of Germany, pp. 179–183

Moore, C. L. and Moore, A. S. (1985) Freestall preferences of dairy heifers for solar heated mats and earth stalls. *Journal of Dairy Science,* **68** (suppl. 1), 238 (abstract)

Muller, L. D., Jaster, E. H., Clark, A. K., Armstrong, D. V. *et al.* (1989) Effect of feeding time and strategy on the time of parturition in dairy cattle. *Journal of Dairy Science,* **71** (in press)

Natzke, R. P., Bray, D. R. and Everett, R. W. (1982) Cow preference for freestall surface material. *Journal of Dairy Science,* **65**, 146–153

Newberry, R. C. and Fisher, L. J. (1988) Do free stall cows prefer sand or rubber tyres. *Hoard's Dairyman,* **131**, 224–225

Owen, J. B. (1979) *Complete Diets for Cattle and Sheep.* Farming Press Ltd, Ipswich, Suffolk, p. 127

Pennington, J. A. and Albright, J. L. (1985) Effect of feeding time, behaviour and environmental factors on the time of calving in dairy cattle. *Journal of Dairy Science,* **68**, 2746–2750

Pennington, J. A., Albright, J. L. and Callahan, C. J. (1985) Sexual activities of Holstein dairy heifers. *Agri-Practice,* **6**, 10–15

Pennington, J. A., Albright, J. L. and Callahan, C. J. (1986) Relationships of sexual activities in estrous cows to different frequencies of observation and pedometer measurement. *Journal of Dairy Science,* **69**, 2925–2934

Plumb, C. S. (1893) Does it pay to shelter milk cows in winter? *Purdue University Agricultural Experiment Station Bulletin,* **47**

Raz, A., Ezra, E. and Ron, M. (1987). In *Israel Cattle Breeder's Association – Israel Holstein Herdbook Annual Report on Productivity and Reproduction,* (ed. M. Mal'an). Hidekel Press, Tel Aviv

Ruckebusch, Y. (1972) The relevance of drowsiness in the circadian cycle of farm animals. *Animal Behaviour,* **20**, 637–643

Schmisseur, W. E., Albright, J. L., Dillon, W. M., Kehrberg, E. W. *et al* (1966) Animal behaviour responses to loose and free stall housing. *Journal of Dairy Science,* **49**, 102–104

Schultz, T. A., Collar, L. S. and Morrison, S. R. (1985) Corral manger misting effects on heat stressed lactating cows. *Journal of Dairy Science,* **68** (suppl. 1), 239 (abstract)

Sniffen, C. J. and Chase, L. E. (1981) Give your cows something to chew on. *Hoard's Dairyman,* **125**, 671

Stephens, D. B. (1980) Stress and its measurement in domestic animals: a review of behavioral and physiological studies under field and laboratory situations. In *Advances in Veterinary Science and Comparative Medicine,* **24**, 179. Academic Press, New York

Tschantz, B. (1978) Reaktionsnormen and Adaption. In *Das Tier im Experiment,* (ed. W. H. Weike. Hans Huber, Bern, p. 33

Whittlestone, W. G., Kilgour, R., De Langen, H. and Duirs, D. (1970) Behavioral stress and the cell count of bovine milk. *Journal of Milk and Food Technology,* **33**, 217–220

Wierenga, H. K. (1984) The social behaviour of dairy cows: some differences between pasture and cubicle system. In *Proceedings of the International Congress on Applied Ethology in Farm Animals.* University of Kiel, Federal Republic of Germany, pp. 135–138

Zeeb, K. and Beilharz, R. G. (1980) In *Applied Ethology and Raising Animals in a Way Which is Suitable for Their Species.* HMSO, London, 406-II, 212–218

Chapter 13

New techniques in calf production

A. J. F. Webster

Although techniques in animal husbandry change and develop, the objectives remain the same. It is important at the outset of any discussion of husbandry systems to remind ourselves of all the major objectives since they are not always compatible. These are:

(1) supply of a high-quality product: a healthy, well-grown calf at 3–6 months of age
(2) maximum efficiency of use of resource: minimizing costs of feed (especially highly expensive milk replacer), housing and management
(3) minimizing losses due to death and disease
(4) ensuring adequate standards of welfare, i.e. provision of an environment to which the calves can adapt without suffering.

In many livestock systems, product quality and food conversion efficiency dominate the economics of the exercise. Uncritical application of this assumption to calf rearing calls for calves to be weaned as young as possible off expensive milk replacer and out of expensive calf houses. Setting aside, for the moment, the question of welfare, this is not necessarily the most economic strategy if it increases losses due to death and disease at the time or impairs subsequent growth.

The most important objective in calf rearing, viewed within the context of economics or welfare, is to keep the calf alive and well. Table 13.1 (abridged from Webster, 1984) illustrates the effects of death and disease in ideal, 'normal' and 'infected' calf units. The 'normal' unit has a mortality rate of 5%, the 'infected' unit

Table 13.1 Economics of death and disease in a calf rearing unit (100 calves)

	Ideal	'Normal'	'Infected'
Calves dying	–	5	21
sick but recovered	–	13	14
Costs (£, 1984) food	3 590	3 492	3 082
veterinary and miscellaneous	300	550	550
calf purchase (£80/head)	8 000	8 000	8 000
sales (£150/head)	15 000	14 250	11 850
Gross profit margin (£)	3 110	2 208	218

From Webster, 1984.

describes the progression of a typical outbreak of disease caused by *Salmonella typhimurium*. The economics are based on 1983 prices, the impact today (1988) is far greater as the value of the newborn calf soars towards £200/head. Even a normal level of disease carried (in 1983) a cost of £900 per batch of 100 calves. A typical outbreak of salmonellosis destroys gross profit altogether.

Bearing this in mind, let me consider a good beef calf purchased in 1988 at a cost of £200. This calf, bucket-fed and early-weaned, may receive 15 kg of milk powder at £850/tonne – an outlay of £12.75. This represents about 0.5% of the nutrients required to bring a beef animal to slaughter weight or a dairy heifer to the point of first calving. Assuming all goes well, this is a satisfactorily modest outlay. However, if the disease status of the unit is less than ideal there may be good economic grounds for devising an alternative weaning strategy even if food costs are substantially increased.

The welfare of calves born into the dairy herd has improved enormously over the last 40 years. It would be nice, albeit self-righteous, to attribute this to an improved awareness of animal rights. In fact, calf mortality statistics (which reflect husbandry practices) are intimately, but inversely, linked to the cash value of calves. When male calves from the dairy herd are in relative surplus and cheap, mortality rates rise towards 20%; when their price exceeds £100, husbandry improves greatly and mortality falls to about 1% (Webster, 1984, p. 3).

Starting calves

This section deals almost entirely with calves from the dairy herd and removed from their mothers shortly after birth. The first essential is to ensure good hygiene at calving since a calf can be infected with, for example, a virulent strain of *Escherichia coli* within the first minutes of life and septicaemia may develop before it acquires adequate protection through colostrum. A new technique for the control of septicaemia and enteritis in calves associated with enterotoxigenic *E. coli* (with K99 antigen) and/or rotavirus is to vaccinate cows 12–14 weeks before calving with a product such as Rotavec K99 (Coopers Animal Health Ltd). This boosts the level of antibody in her colostrum and milk. The importance of conferring protection through colostral antibodies absorbed across the gut wall during the first day of life is well known. There is also good evidence that a continuing supply of antibody in milk can sustain local protection at the mucosal surface of the gut against localized infection with, for example, rotavirus (Snodgrass *et al.*, 1980, 1982).

This approach is obviously most applicable to suckler herds where calves drink from their mothers, but intake is uncontrolled and often rather generous and hygiene at calving may be less than ideal, particularly for herds that calve indoors. The approach may be applied to dairy herds if a problem exists; in which case it should be combined with a feeding system based on stored colostrum and whole milk up to weaning.

A calf weighing 45 kg at birth has an abomasal capacity of about 1.5 litres. Individual feeds, whether from teat or bucket, should be restricted to this amount for the first week of life to avoid overloading the abomasum and so permitting undigested food and live microorganisms to enter the small intestine. If calves are removed from their mothers shortly after birth, attempts should be made for them to drink 1.5 litres colostrum on at least three occasions during the first 24 h (and preferably more).

Bought-in calves

Many specialist units buy in calves knowing neither their provenance nor their colostral status. It is possible to assess the antibody status of calves, either by direct assay for immunoglobulin G (IgG) or from the indirect zinc sulphate turbidity test (Penhale *et al.,* 1970). While it is undoubtedly true that calves with little or no colostral-derived immunity are highly vulnerable to infection, the correlation between antibody levels and subsequent calf viability tends to be low within a normal population of calves that have received *some* colostrum (Barber, 1978). There is a new, commercial latex agglutination test to determine the IgG_1 status of calves (White, 1986). However, work by Simmons and Hoinville (personal communication) suggests that many calves failing this test had adequate IgG_1 levels (25.4 ± 8.2 mg/ml serum). Tests for IgG status are a useful way of ensuring that suppliers give their calves enough colostrum; they should not however be interpreted too rigidly.

Ideally, rearers should pick up calves directly from a limited number of farms where the husbandry is known. Unfortunately this is seldom possible, partly because it conflicts with another central precept of good hygiene, namely the 'all in – all out' policy. Thus calves tend to be transported to market at about one week of age, bought by dealers and held on their premises for varying periods of time before being returned to market, or preferably transported directly to the specialist rearing unit. Whatever are one's views of the ethics of this arrangement, it undoubtedly maximizes the opportunity for cross infection, especially with *Salmonella typhimurium*, at a time when the general resistance of the calf is low. The rearer who buys in his calves through a dealer would be wise to assume that at least some are infected with salmonella and/or other potential pathogens and institute an appropriate course of prophylaxis. Our calves bought from dealers are routinely given a course of potentiated sulphonamides active against most (but not all) strains of salmonella, and also an injection of vitamins A and D. Colostrum is the natural source of vitamin A (especially) to the neonatal calf. Calves that are deprived of colostrum or born in late winter when the vitamin A status of their dam is low, require vitamin A supplementation primarily to ensure the integrity of epithelial cell turnover and so reduce the risk of epithelial disorders such as enteritis or pneumonia.

Simple infection is not the only problem for bought-in calves. They may also be suffering from any combination of the following: cold, hunger, dehydration and exhaustion. On arrival they should be rested for at least 3 h in the comfort and warmth of deep, clean bedding. There is no absolute need to install newly-delivered calves immediately into individual pens since any infectious organisms will have been spread already. Moreover, ventilation is not a major issue at this stage. It is a good policy to give the calves 2×1.5 litres of electrolyte/glucose-glycine solution in the first 24 h rather than milk replacer to reduce the risk of enteritis. Many calf rearers however successfully feed 1.5 litres of milk replacer at 125 g/l from the outset.

Feeding calves to weaning

In the UK, 'normal' procedure is to wean calves off milk as soon as possible, which has come to mean about 5 weeks of age. In many other developed countries with equally intensive agriculture, weaning may be at 8–12 weeks. There are several good reasons for early weaning:

(1) cost of milk replacer
(2) cost of specialist accommodation for milk-fed calves
(3) labour cost of (especially) bucket-feeding systems
(4) problems of diarrhoea in milk-fed calves
(5) problems of pneumonia in densely stocked calves.

Taken together, these factors make a strong case for weaning calves as soon as possible and turning them out into the fresh air in low cost, naturally-ventilated 'follow-on' units. However, none of these factors on its own is sufficient reason for very early weaning.

Large numbers of new systems for rearing calves have been tried out in recent years. These have been made possible through improvements in technology relating to calf feeds and feeding machines.

Composition of milk replacers

Conventional milk replacers, sometimes called 'sweet' or 'high-fat' powders are made by adding animal fat to skim-milk. The expression 'high-fat' is a misnomer, since fat content is similar to, or a little less than that of whole milk (300 g/kg dry matter (DM)). It merely means higher than formerly when emulsification techniques were less developed and fat content was as low as 50 g/kg DM. These conventional 'sweet' milk replacers are physiologically correct for the immature digestive system of the calf, nutritionally balanced, and form a clot in the abomasum providing a balanced release of nutrients over a period of 8 h or more. They are particularly suitable for once-daily feeding but, of course, tend to sour rather quickly.

While the Common Agriculture Policy within the EEC continues to subsidize the cost of skim milk, these conventional powders will continue to be favoured by calf rearers. It is however possible to rear calves on 'milkless' powders in which both butterfat and casein have been replaced by alternatives. The main alternatives to casein are the whey proteins (albumen and globulins) and high-quality vegetable proteins such as soya appropriately (and expensively) treated to destroy antigenic compounds (Kilshaw and Sissons, 1979). These milkless powders, lacking casein, do not form a clot in the abomasum. They are usually acidified to pH 4.2 which not only preserves them outside the calf but also keeps abomasal pH below 4 and so ensures the destruction of microbes before entry to the small intestine (Webster, 1984, p. 68).

The digestibility of milkless milk replacers is lower and slower than that of whole milk or sweet powders. A few calves find them unpalatable. However, the better acidified milkless powders have not been associated with any increase in enteric problems (Thickett et al., 1983). At present they are not cheap enough relative to sweet powders to catch on but if and when the subsidy on skim milk disappears their popularity is likely to increase.

The most effective way to minimize the risk of orally ingested pathogens entering the small intestine and causing enteritis is to ensure that all food enters a properly functioning rumen and is degraded by the normal, symbiotic population of microorganisms therein. This is, of course, why enteric problems in calves tend to disappear after weaning. In fashionable terms, the rumen population may be described as the ultimate probiotic – broad spectrum, effective and free. There are conceptual reasons why commercial probiotics for calves containing, for example lactobacillus species may reduce the risk of enteric infection when included in milk powders and taken direct into the abomasum. However, the published work on probiotics in the simple-stomached pig is not very encouraging (Pollman, 1986). Nearly all the unpublished but properly conducted trials with calves that I have been privileged to see are equally unconvincing. Often success has been claimed on commercial units where calves are also receiving furazolidone and tetracycline. Such claims are worthless. There is no logical justification for including probiotics in diets for calves after weaning is complete and the rumen is working properly.

Feeding systems

Table 13.2 compares different feeding systems for calves. The figures are from Liscombe Experimental Husbandry Farm (1981) but with 1988 prices for the UK added.

Table 13.2 Comparison of calf-feeding systems

Rearing method	Days to weaning	12-week weight (kg)	Feed (kg)			Food costs £, 0–12 weeks	p/kg gain
			Milk powder	Starter pellets	Hay		
Bucket feeding							
twice-daily, warm	34	100	14	118	12	33.98	58
once-daily, warm	34	95	15	108	12	33.03	62
once-daily, cold	45	90	15	112	12	33.75	70
Teat feeding							
automatic dispenser, warm	34	106	30	100	12	44.34	69
mild acid powder, cold	35	105	29	105	12	44.39	70

All calves were purchased weighing 42 kg on average and weaned at approximately 60 kg liveweight. Data from Liscombe (1981) with 1988 prices (£/tonne) namely, milk powder 850; starter pellets 180; hay 70.

The two teat feeding systems describe calves reared in groups to weaning and with milk replacer available *ad libitum* either at 40°C from an automatic dispenser or at ambient temperature. In the latter case the skim-based powder was mildly acidified to pH 5.7. Rearing calves in groups with unrestricted access to milk currently costs about £11 more than bucket feeding to produce a slightly stronger calf at 12 weeks of age. Twice-daily bucket feeding was the most successful system, considered strictly in terms of feed costs. Once daily feeding of cold milk was not a success; it clearly exceeded the capacity of the calves to adapt.

Calves may be weaned off milk when eating 1 kg/day of a dry, cereal-based starter ration. This provides about 12 MJ metabolizable energy which is about 10% above the maintenance requirement of a 60 kg calf. Weaning calves when intake of starter is below 1 kg/day will not only cause a set-back in growth but may predispose to infections post-weaning, especially pneumonia.

One of the problems of teat-feeding systems is that intake of starter is likely to be well below 1 kg/day by 5 weeks of age (Figure 13.1). Sudden weaning of such calves

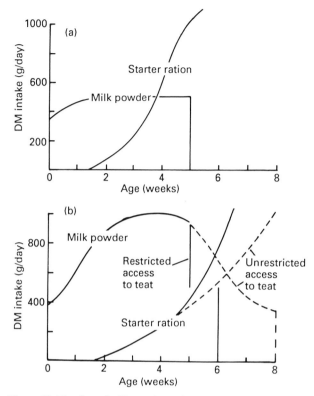

Figure 13.1 Intakes of milk powder and starter rations (*a*) calves bucket-fed to 5 weeks, (*b*) unrestricted teat feeding to 5 weeks followed by restricted access to 6 weeks or unrestricted access to 8 weeks (Webster, 1984). DM, dry matter

can cause a severe set-back, more than wiping out any weight advantage achieved through previous unrestricted access to milk powder. When access to milk remains unrestricted most calves will consume 1 kg/day of starter by 8 weeks of age and this approach may suit some farmers. Alternatively, access to milk can be reduced progressively from 4 weeks. The easiest way to do this is to apply a screw clamp to the milk line and tighten it progressively. Calves find they have to work harder for their milk and reduce intake accordingly.

There is no best combination of feed and feeding system for rearing calves to weaning. The bucket-based systems have the advantage of low feed cost, good hygiene and early recognition of disease, since loss of appetite is usually the first clinical sign. Systems in which calves are reared in groups and drink milk from teats have the advantage of convenience. Furthermore, they are quite as healthy as bucket systems with individual penning (Thickett *et al.*, 1983). Group rearing from a communal teat potentially increases the risk of spreading disease by contagion. However, this is rarely a problem in practice, partly because 'normal' sucking of milk through teats as required provides good digestion in the abomasum and reduces the risk of pathogens entering the small intestine. Moreover, Wathes (1988) has recently demonstrated at Bristol that the enteropathogens *E. coli* and *Salmonella typhimurium* are equally infective whether acquired by inhalation or by contagion through the oral route. This observation seriously weakens the case for individual penning.

The best system for any farmer will depend on factors such as the quality, origin and disease status of his calves, the type of buildings and labour available to him and what he intends to do with his animals, principally whether or not he intends to sell them early. These questions usually justify good specialist advice tailored to the individual unit.

Calf housing and respiratory disease

Environmental needs and housing design for calves have been recently considered in detail (Webster, 1985, Ch. 4, 5). Here I wish to consider only recent developments in our understanding of the effect of the environment on respiratory disease in calves, in particular pneumonia occurring typically in animals 3–8 weeks of age.

The aetiology of calf pneumonia is complex. Thomas *et al.* (1982) identified five viruses, four species of mycoplasma and 19 bacteria from eight outbreaks of calf pneumonia in the UK. There are undoubtedly more. Multivalent vaccines have been used with partial success to control some outbreaks of pneumonia (Howard *et al.*, 1987) but this approach is never likely to be the complete answer partly because of the multiplicity of organisms involved and partly because many of these organisms can inhabit the respiratory tract without causing disease. They can and do exert a vaccinal effect, i.e. stimulate the immune system. Despite this an animal may succumb to an acute attack of pneumonia associated with an organism that it may have been harbouring for weeks.

Pneumonia may thus be seen as a failure of homeostasis which occurs when the combined challenge from potential pathogens and other aerosols harmful to vulnerable epithelial surfaces in the lung exceeds the animal's capacity to clear them by mechanical removal up the mucociliary escalator or by immune mechanisms for killing and removal (e.g. phagocytes). The housing environment may affect this balance via:

(1) the concentration in inspired air of pathogens and noxious aerosols
(2) sites of deposition of aerosols in the lung and their clearance by the mucociliary escalator
(3) the capacity of the animal to mount an immune response.

The concentration (C_b, n/cm^3) of any aerosol in the air of a livestock building is determined by its rate of release (R, n/cm^3 per h), entry from incoming air (C_i, q_v) and clearance (q/h) by pathways such as ventilation (q_v), death *in situ* (q_d), etc. (q_e) (Wathes, Jones and Webster, 1983).

At equilibrium

$$C_b = 1/q \ (R + C_i . q_v)$$

Most bacteria and viruses arise from the animals themselves. Other aerosols such as fungal spores and gaseous pollutants arise mainly from feed, bedding and excreta. The effect of a gaseous pollutant such as ammonia is obviously proportional to its concentration in air. Equally, fungal spores (e.g. aspergillus species and the thermophilic actinomycetes responsible for 'farmer's lung' in man and stabled animals) exert an effect proportional to their concentration in air since

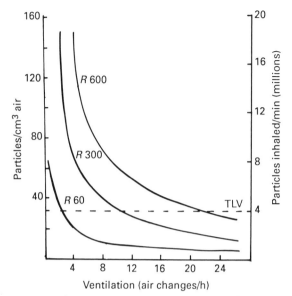

Figure 13.2 Effects of release rate (R, particles/cm^2 per h) and ventilation rate (q_v/h) on the concentration of respirable particles in stable air and inhaled by a 500 kg horse (Webster *et al.*, 1987)

they have a relatively long survival time in air and are antigenic whether dead or alive (Clarke, 1987).

Ventilation is the prime route for clearance of spores and other components of total respirable dust (particles less than 5 µm in diameter). Figure 13.2 (Webster *et al.*, 1987) illustrates effects of varying R and q_v on the concentration of respirable particles in a stable. Ventilation improves the 'cleanliness' of the air in linear fashion. 'Dirtiness' (C_b) is therefore proportional to $1/q_v$ and increases very rapidly as q_v falls below four air changes per hour. When R is low (e.g. 60) as in a well-managed equine stable, increasing q_v above 6/h confers little further advantage. In calf houses with bedding it is unrealistic to expect R to be less than 300.

The situation is quite different for the viruses and bacteria which are the primary and secondary pathogens in pneumonia. Most of these die within a few seconds of exposure to air (Donaldson, 1978), although the few (perhaps 1%) that survive the initial process of aerosolization may survive for minutes or hours (Cox, 1987). Temperature and humidity affect q_d but these effects are relatively minor (Donaldson, 1978; Wathes, Jones and Webster, 1983). For most pathogens of the respiratory tract therefore, death rate, (q_d) values are 30 or higher. In a poorly ventilated calf house where $q_v = 3$, it follows that ventilation contributes 10% or less to total clearance of pathogens.

Since infected animals are the usual source of pathogens, inhaled dose increases in direct proportion to stocking density but is relatively unaffected by ventilation. Figure 13.3 (Webster, 1984, p. 90) illustrates this point. Increasing space allowance from 5 to 10 m^3 per calf halves the concentration of airborne bacteria; increasing ventilation rate from 4 to 40 air changes per hour reduces it by only one-third.

Ventilation thus has relatively little direct effect on the concentration of primary pathogens (R_p). Doubling stocking density doubles R_p. The amount of ventilation

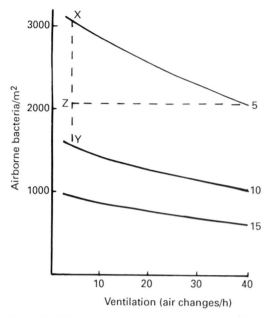

Figure 13.3 Effect of space allowance (5, 10 or 15 m^3/calf) and ventilation rate on the concentration of airborne bacteria in a calf house. The intervals XY and XZ indicate respectively the effects of increasing space allowance from 5 to 10 m^3/calf (XY) and increasing ventilation rate from 4 to 40 air changes/h (XZ)

required to return R_p to its original value is very large ($q_v > 60$/h), uncomfortably draughty to the point of causing cold stress and quite impractical.

One should not conclude from this argument that ventilation is unimportant. It removes fungal spores and other respirable pollutants which may compromise the mechanical and immune defences of the respiratory tract and so reduce resistance to infection. It also controls humidity within the animal house and thereby the survival time of microbes in air (Webster, 1984, p. 108). In a well-designed, properly stocked calf house where R is low, satisfactory air hygiene can be achieved at ventilation rates no greater than six to eight air changes per hour which can readily be achieved by natural ventilation without causing draughts (Mitchell, 1978).

Table 13.3 gives my recommended values for stocking density and ventilation rate in calf houses. These may appear generous but they are necessary to maintain

Table 13.3 Air space and ventilation requirements for calf houses

	Pre-weaning (0–6 weeks)	Post-weaning (6–12 weeks)	Veal calves (to 200 kg)
Minimum air space per calf (m^3)	6	10	15
Ventilation rate, air changes/h			
winter minimum	4	6	6
summer maximum	12	18	18

the concentration of respirable pathogens and other pollutants within acceptable levels in circumstances where infection is endemic. Building design cannot eliminate infection in these circumstances. Equally, over-stocking and poor ventilation do not *cause* pneumonia in the absence of a primary pathogen, which is why specific pathogen-free pigs and poultry can thrive in environments that are heavily polluted by the standards of a calf house.

In designing calf houses it is far more important to reduce the challenge from primary pathogens than it is to attempt to avoid stresses such as cold that may reduce host resistance. Calf rearers in the northern USA and Canada have provided ample practical proof of this, rearing calves out of doors in individual hutches at air temperatures that may, on occasions, fall below −20°C. When the concentration of pathogens is negligible there is no disease. Moreover, growth rate and food conversion efficiency are usually similar to those of healthy calves reared indoors in a thermoneutral environment (*see* Chapter 12).

In the UK, so long as calves are provided with shelter, suitable bedding and are not exposed to draughts, they are only likely to feel cold if deprived of food (as occurs during transit) or if they are fevered. Such animals will, on occasions, appreciate some supplementary heating.

Most cases of calf pneumonia occur shortly after weaning. Similarly, pasteurella pneumonia ('shipping fever') in beef cattle is clearly associated with a radical change in environment and feeding. Martin *et al.* (1981) have conducted a most powerful epidemiological study of the aetiology of pneumonic pasteurellosis in beef cattle. There has been no epidemiological study on calf pneumonia in the UK that compares in scope with Martin's 'Bruce County' study. However, despite the lack of hard evidence, it is difficult to avoid the conclusion that calf pneumonia may be precipitated by the act of weaning. Some of the possible reasons are fairly obvious, e.g. mixing of calves and movement to a new environment, often containing older, recovered animals still carrying infection. On clinical (if not yet scientific) grounds it is equally reasonable to assume that following sudden weaning, calves may become undernourished leading to immunosuppression (Kelley, 1980). Moreover, the study by Martin *et al.* (1981) would suggest that problems of rumen indigestion associated with excessive consumption of concentrate foods may lead to the absorption of toxins which compromise the defences of the lung, Most of the herd problems of calf pneumonia that I see at present are second opinion cases where the disease is chronic and has not responded satisfactorily to vaccination therapy or alterations to building design. In these circumstances a change in weaning strategy often works. For example, calves may be bought in, in groups of 30, reared together on an automatic milk dispenser in simple 'follow-on' type accommodation from the time of arrival, weaned gradually (by constricting the milk supply) over 2–3 weeks and kept in the same building until 'turn out' in the spring. This effectively eliminates most managemental and microbiological disturbances. Feed cost per calf may be increased by £15, but if the lives of two calves are saved thereby the farmer will still be better off.

Contract calf rearing

Problems of infectious disease in calves taken from their farm of origin and reared on specialist units tend to be most severe when buildings are occupied continuously and by cattle of different ages. The incidence of pneumonia in particular is high when weaned calves constantly enter an air space occupied by older animals which may appear clinically normal but still carry pathogens such as respiratory syncytial virus (Thomas, 1978). This problem has been the main reason for the emergence of specialist units contracting to supply calves at an age of about 12 weeks and a weight (depending on breed) of 110–140 kg. Most of these calves are destined for intensive or semi-intensive beef systems. The objective therefore is to ensure that they leave the contract calf rearing unit preconditioned to a high concentrate feed and free of infectious disease. Table 13.4 presents 1986 performance and economic figures for contract calf rearing units recorded by the Meat and Livestock Commission (MLC, 1987). The top third achieved a gross profit margin of £34 per head. The £18 difference in economic return between the top and bottom thirds can be attributed largely to a reduction in cost of milk powder and the virtual elimination of death and disease, achieved in part by buying in heavier, stronger, more expensive calves.

Table 13.4 Performance of the top third and bottom third specialist contract calf rearing units recorded by the Meat and Livestock Commission (MLC, 1987)

	Bottom third	Top third
Performance		
Weight (kg) at start	49	54
at sale	111	117
Feeding period (day)	90	84
Daily gain (kg)	0.69	0.78
Mortality (%)	5	0
Feeds (kg), milk powder	19	10
concentrate	177	169
Financial results (£/head)		
Calf sales	193	215
Less calf cost, mortality	127	140
Variable costs, milk powder	14	8
concentrates	28	26
other	8	7
Gross margin per head (£)	16	34

There is no doubt that contract calf rearing can be a highly successful enterprise despite the fact that satisfactory gross profit margins can only be achieved by restricting milk powder intake to the absolute minimum. It is also a system that can go quite catastrophically wrong.

The main disease problems are those of bloat, acidosis and urolithiasis, all of which can be linked to high intakes of high starch, low fibre concentrate feeds. A conventional starter concentrate ration for calves has the following formulation: crude protein 16–18%, crude fibre 6.5%, oil 3.5%. It will be based predominantly on cereals, such as rolled oats, barley and flaked maize and protein-rich byproducts such as soya bean meal. These ingredients are excellent sources of nutrients for a young animal with a high growth potential. Autumn-born calves reared conventionally over winter to a target weight of 180–200 kg when turned out to grass in the spring at about 6 months of age should be given no more than 3 kg/day of this highly concentrated starter ration plus *ad libitum* access to good hay or silage. This usually ensures a satisfactory balance between starch and digestible fibre and so permits normal healthy development of the rumen. However, contract calf rearers aim for the maximum possible weight gain by 12 weeks of age, the earliest age at which calves may safely enter an intensive beef unit stocked with animals up to slaughter weight. To achieve this, calves are usually given a *ad libitum* access to a conventional starter ration (crude fibre 6.5%) plus hay or barley straw. When all goes well this achieves performance like that of the top third in Table 13.4.

However, unrestricted access to starchy concentrates creates a risk of promoting excessively rapid, unstable rumen fermentation leading to acidosis and bloat. The risk increases if the forage on offer is unpalatable or if there are management faults such as irregular feeding, dirty water bowls, etc.

Strategies designed to reduce ruminal acidosis and bloat include the addition of sodium bicarbonate in the starter ration (Kellaway *et al.*, 1977) or the incorporation of 180 g/kg chopped straw into a coarse pellet (Thomas and Hinks, 1982). Neither approach offers a complete solution. Sodium bicarbonate buffers the consequences of rapid starch fermentation but has no direct effect on fermentation rate. Straw chopped finely enough to be incorporated into a pellet acts as a poorly digestible material which dilutes the fermentable substrates in the rumen but does not stimulate rumination to the same extent as forage having a fibre length greater than 20 mm. Increasing rumination time is the most effective and cheapest way of increasing the flow into the rumen of saliva which both dilutes and buffers the rumen contents. In a 12-week-old calf eating a ration containing sufficient long fibre to stimulate rumination for 8 h/day, the normal rate of physiological secretion of sodium bicarbonate into the rumen probably exceeds 200 g/day (Kay and Pfeffer, 1970). When rumination time is 2 h or less, it may be as low as 50 g/day. Supplementation of starter pellets with sodium bicarbonate is usually at concentrations within the range 15–35 g/kg. This would achieve an intake of 60–140 g/day bringing sodium bicarbonate entry into the rumen up to that of a calf ruminating 'normally'. Salivary sodium bicarbonate is, however, provided free of charge since bicarbonate is an inevitable waste product of energy metabolism and the sodium is recycled.

Young calves on high concentrate diets also run an increased risk of urolithiasis (urinary calculi). This can be a particularly severe problem in male calves because of the tortuous passage of the urethra around the sigmoid flexure of the penis. Salts, mostly phosphates such as magnesium ammonium phosphate, crystallize and expand in the urethra, restricting urine flow and frequently causing total blockage, severe pain and death. Factors which may precipitate urolithiasis in calves on high concentrate diets include the following:

(1) diets which are naturally rich in phosphorus and magnesium (over 60 g/kg) and where calcium:phosphorus (Ca:P) ratio is less than 1.5:1 (Rice and McMurray, 1981)
(2) low water intake leading to an increased concentration of all salts in the urine
(3) secretion of mucus into the urine which forms a medium on which salts may crystallize
(4) low output of faecal dry matter which reduces faecal excretion of phosphates and increases excretion via the urine (Scott and Buchan, 1988).

Preventive measures against urolithiasis include (ADAS, 1988):

(1) increasing Ca:P ratio to 2:1 when natural levels of P and Mg are high
(2) addition of 5–10 g/kg ammonium chloride to prevent the formation of alkaline urine. This is however most unlikely in calves on high cereal diets
(3) increasing urine volume by inclusion of sodium chloride (NaCl) at concentrations of at least 10 g/kg. Higher concentrations are inclined to reduce intake. If sodium chloride is added, water must be freely available.

Most of the preventive measures for bloat, acidosis and urolithiasis are designed to treat conditions that need not have arisen in the first place if the starter diet had been better suited to the normal processes of ruminant digestion. There is a real need to develop and market for contract calf units improved complete diets that include less starch than at present and more fibre of a physical form long enough to stimulate normal rumination and salivation and a chemical form sufficiently digestible to promote high food intake and acceptable weight gains. Alkali-treated straw is one option but there are many other sources of digestible fibre which are currently used with success in diets for lactating dairy cows (Visser and Steg, 1988) and which could be incorporated with equal success into complete diets for young calves.

Improved husbandry systems for veal calves

The conventional system of confining veal calves in individual wooden crates and feeding them twice daily from buckets on a liquid diet of milk replacer has received much criticism on welfare grounds. The most obvious insult to the welfare of the calf is to confine it in a box where, in the latter stages of growth, it cannot turn round, groom itself properly, adopt a normal sleeping position, or even stand up and lie down without difficulty.

A more searching examination of a typical conventional commercial veal unit reveals abuses to each of the five freedoms from hunger and malnutrition, thermal and physical discomfort, injury, suppression of 'normal' behaviour, fear and disease (Webster, 1984; Webster, Saville and Welchman, 1986):

(1) the calves are deprived of dry feed. This completely distorts the normal development of the rumen and encourages the development of hair balls which may lead to chronic indigestion
(2) intake of dietary iron is severely restricted to ensure 'white' meat
(3) wooden crates with slatted floors are uncomfortable and severely restrict behaviour
(4) young calves are seldom cold in veal units but well-grown calves frequently experience acute heat stress
(5) the incidence of infectious disease is high and often kept under control only by liberal and repeated administration of antibiotics
(6) deprived of solid food to eat and ruminate upon, veal calves in crates engage in various purposeless oral activities such as crate licking or tongue rolling. These forms of stereotyped behaviour are usually, although not entirely convincingly, taken as indicators of stress (Dawkins, 1980).

In recent years, commercial producers in several countries have been rearing calves for veal in groups in yards bedded with straw and wood shavings. The calves are given unrestricted access to milk replacer dispensed through teats at blood heat. In the UK this has become known as the Quantock system after its pioneers, Quantock Veal Ltd. It has the clear merit that the calves are permitted free expression of most normal activities. Access to straw also gives them the opportunity to eat something solid, although it provides little in the way of nutrients. Iron is added to the milk powder at about 25 mg/kg which is a little higher than in most conventional units in Europe but still low enough to keep the meat pale pink. Although the system has had some success in the UK, it has not proved commercially attractive in countries with a large conventional veal industry due in part to innate conservatism and to the amount of capital invested in the conventional system. However, it is also argued that calves reared in groups and given unrestricted access to milk powder cannot compete economically with those in crates due to poorer food conversion efficiency and a high incidence of ill-health. A greater variability in growth rate and meat colour also tends to create a less uniform product. The main drawback of the Quantock system, group rearing calves with *ad libitum* access to milk powder from teats, is that it is uneconomic (Table 13.5). In our experience, one tonne of milk replacer reared five calves in crates but only four on the Quantock system (Webster, Saville and Welchman, 1986).

Table 13.5 Performance of Friesian/Holstein male calves

| | Rearing system | | | | SED |
| | Boxes | | Group reared | | |
	Milk only	Milk + dry food	Bucket-fed	Teats ad libitum	
Numbers reared	15	47	18	23	
Milk powder consumption (kg)	202	196	200	247	
Dry food consumption (kg)	nil	27	unknown		
Weight at slaughter (kg)	175	188	189	194	4.6
Liveweight gain (kg/day)	1.10	1.20	1.15	1.27	0.04
Food conversion ratio*	1.63	1.50	1.41	1.70	0.05
Killing-out (%)	58.1	56.8	55.7	58.3	0.06

Figures are overall mean values with standard errors of differences between means (SED).

* (kg milk powder + 0.6 × kg dry food) per kg liveweight gain.

Our most successful approach to rearing veal calves has been the 'Access' system (Table 13.6). Here calves with transponders have access to a computer controlled feeder from which they receive controlled amounts of milk powder plus about 300 g/day of a palatable unmilled starter ration containing sufficient fibre to promote normal ruminal development. This system was not only as efficient as the conventional buckets and boxes approach in terms of food conversion efficiency in healthy calves but greatly reduced losses due to death and disease, making it the only system in our hands to generate a satisfactory gross profit margin (Table 13.7). We also concluded that it is impossible to guarantee that 'white' veal can be produced without creating clinical anaemia in some calves (Webster, Saville and Welchman, 1986; Welchman, Whelehan and Webster, 1988), a conclusion at

Table 13.6 Performance of Hereford × Friesian female calves

	Group-rearing system		SED
	Quantock	Access	
Numbers reared	60	28	
Milk powder consumption (kg)	194	178	
Starter feed consumption (kg)	nil	30	
Weight at slaughter (kg)	162	168	3.4
Liveweight gain (kg/day)	1.09	1.12	0.04
Food conversion ratio*	1.63	1.52	0.05
Killing-out (%)	57.2	57.0	0.04

Figures are overall mean values with standard errors of differences between means (SED).

* (kg milk powder + 0.6 × kg starter ration) per kg liveweight gain.

Table 13.7 Comparative economics of different rearing systems for veal calves

	Friesian/Holstein bulls		Hereford × Friesian heifers	
	Boxes	Quantock	Quantock	Access
£ per calf purchased				
Calf price	70	70	50	50
Milk powder	121	148	116	107
Dry food	–	–	–	5
Straw etc.	–	10	10	10
Total variable costs	191	228	176	172
Selling price	225	250	200	209
Gross profit (ideal)	+34	+22	+24	+37
Less deaths/disease	30	30	30	3
	+3	−8	−6	+34

variance with an earlier recommendation based on a much less intensive study (Bremner *et al.*, 1976).

Our studies of the performance, health and behaviour of veal calves provoked the following conclusions:

(1) if calves are reared for veal in individual compartments then their internal width should be not less than 900 mm for any calf weighing over 100 kg
(2) no calf should be deprived access to palatable, digestible dry food after 2 weeks of age
(3) diets for veal calves should contain sufficient iron to maintain sound health and vigour.

The first two conclusions are the subject of current recommendations for legislation in the UK. The third will be more difficult to achieve. However, diets that are adequate in iron only make the meat slightly pinker. It may be that for every gourmet who insists that his veal should be 'white' there are many more potential consumers who would eat veal if they approved of the method of production.

Conclusions

Most of the problems of calf rearing outlined in this review can be attributed to the management system that man has imposed, more or less recently, on the animals. In my opinion, there are three main aspects of calf rearing that stand out as being unsatisfactory at present but amenable to improvement in the light of existing knowledge.

Marketing

The commercial practice of moving beef calves from the dairy herd through one to three markets in the first month of life evolved at a time when such calves had very little value and high losses were acceptable. This strategy is entirely inappropriate today in terms both of economics and welfare. It is in the interests of the industry to ensure that these calves are moved as little as possible and are managed so as to minimize infection and digestive upsets and maximize immune resistance.

Contract calf rearing

This system has evolved largely to offset the insults imposed by a faulty marketing strategy. On balance it does more good than harm. However, the desire to provide as much weight gain as possible by 12 weeks of age is driving calf rearers to feed high-risk diets. There is a need to devise better complete diets for these calves that will improve health and performance in the long term even if 12-week weights are reduced slightly.

Veal calves

Calves can be reared for quality veal on diets that permit normal rumen development and in housing that permits normal expression of behaviour. Such animals are usually healthier than conventionally boxed veal calves and can compete economically. However, it is not possible to ensure 'white' veal without inducing clinical anaemia in some individuals. Should the visual appearance of meat assume preference over the health and vigour of the animals? I think not.

References

Agricultural Development and Advisory Service (ADAS) (1988) ADAS Technical News, Nutrition Chemistry

Barber, D. M. L. (1978) Serum immune globulin status of purchased calves; an unreliable guide to viability and performance. *Veterinary Record*, **120**, 418–420

Bremner, I., Brockway, J. M., Donnelly, H. T. and Webster, A. J. F. (1976) Anaemia and veal calf production. *Veterinary Record*, **92**, 203–205

Clarke, A. F. (1987) A review of environmental and host factors in relation to equine respiratory disease. *Equine Veterinary Journal*, **19**, 435–441

Cox, C. S. (1987) *The Aerobiology Pathway of Microorganisms.* John Wiley, Chichester

Dawkins, M. (1980) *Animal suffering; the Science of Animal Welfare.* Chapman and Hall, London

Donaldson, A. I. (1978) Factors influencing the dispersal, survival and deposition of airborne pathogens of farm animals. *Veterinary Bulletin*, **48**, 83–94

Howard, C. J., Stott, E. J., Thomas, L. H., Gourlay, R. N. *et al.* (1987) Protection against respiratory disease in calves induced by vaccines containing respiratory syncytial virus, parainfluenza type 3 virus, *Mycoplasma bovis* and *M.dispar*. *Veterinary Record*, **121**, 372–376

Kay, R. N. B. and Pfeffer, E. (1970) Movements of water and electrolytes into and from the intestine of the sheep. In *Physiology of Digestion and Metabolism in the Ruminant*, (ed. A. J. Phillipson). Oriel, Newcastle-on-Tyne, pp. 390–402

Kellaway, R. C., Thomson, D. J., Beever, D. E. and Osbourn, D. F. (1977) Effects of NaCl and NaHCO₃ on food intake, growth rate and acid-base balance in calves. *Journal of Agricultural Science*, **88**, 1–9

Kelley, K. W. (1980) Stress and immune function: a bibliographic review. *Annales Recherche Veterinaire*, **11**, 445–478

Kilshaw, P. J. and Sissons, J. W. (1979) Gastrointestinal allergy to soyabean protein in pre-ruminant calves. Antibody production and digestive disturbances in calves fed heated soyabean flour. *Research in Veterinary Science*, **27**, 361–365

Liscombe Experimental Husbandry Farm (1981) *Beef bulletin No. 1. Calf Rearing*. Ministry of Agriculture, Fisheries and Food, UK

Martin, S. W., Meek, A. H., Davis, D. G., Johnson, J. A. *et al.* (1981) Factors associated with morbidity and mortality in feedlot calves. The Bruce County Beef Project, Year Two. *Canadian Journal of Comparative Medicine*, **45**, 103–112

Meat and Livestock Commission (MLC) (1987) *Beef Yearbook 1986*. MLC, Bletchley

Mitchell, C. D. (1978) *Calf Housing Handbook*, 2nd edn. Scottish Farm Buildings Investigation Unit, Scottapress, Aberdeen

Penhale, W. J., Christie, G., McEwan, A. D., Fisher, E. W. *et al.* (1970) Quantitative studies on bovine immunoglobulins. *British Veterinary Journal*, **126**, 30–36

Pollman, D. S. (1986) Probiotics in pig diets. In *Recent Advances in Animal Nutrition – 1986*, (ed. W. Haresign and D. J. A. Cole). Butterworths, London, pp. 193–205

Rice, D. A. and McMurray, C. H. (1981) Urolithiasis in calves. *Veterinary Record*, **109**, 88

Scott, D. and Buchan, W. (1988) The effects of feeding pelleted diets made from either coarsely or finely ground hay on phosphorus balance and on the partition of phosphorus excretion between urine and faeces in the sheep. *Quarterly Journal of Experimental Physiology*, **73**, 315–322

Snodgrass, D. R., Fahey, K. J., Wells, P. W., Campbell, I. *et al.* (1980) Passive immunity in calf rotavirus infections. Maternal vaccination increases and prolongs immunoglobulin G₁ antibody secretion in milk. *Infection and Immunity*, **28**, 344–349

Snodgrass, D. R., Nagy, L. K., Sherwood, D. and Campbell, I. (1982) Passive immunity in calf diarrhoea. Vaccination with K99 antigen of enterotoxigenic *Escherichia coli* and rotavirus. *Infection and Immunity*, **37**, 586–591

Thickett, W. S., Cuthbert, N. H., Brigstocke, T. D. A., Lindeman, M. A. *et al.* (1983) A note on the performance of and management of calves raised on cold acidified milk replacer fed a*d libitum*. *Animal Production*, **36**, 147–150

Thomas, D. B. and Hinks, C. E. (1982) The effect of changing the physical form of roughage on the performance of the early-weaned calf. *Animal Production*, **35**, 375–384

Thomas, L. H. (1978) Disease incidence and epidemiology – the situation in the UK. In *Respiratory Diseases in Cattle*, (ed. W. B. Martin). Martinus Nijhoff, The Hague, pp. 57–65

Thomas, L. H., Gourley, R. N., Stott, E. J., Howard, C. J. *et al.* (1982) A search for new microorganisms in calf pneumonia by the inoculation of gnotobiotic calves. *Research in Veterinary Science*, **33**, 170–182

Visser, H. de and Steg, A. (1988) Utilisation of by-products for dairy cow feeds. In *Nutrition and Lactation in the Dairy Cow*, (ed. P. C. Garnsworthy). Butterworths, London, pp. 378–394

Wathes, C. M. (1988) Airborne transmission of enteric pathogens in farm livestock. In *Environment and Animal Health*, (ed. I. Ekesbo). Swedish University of Agricultural Science, Skara, Sweden, pp. 421–427

Wathes, C. M., Jones, C. D. R. and Webster, A. J. F. (1983) Ventilation, air hygiene and animal health. *Veterinary Record*, **113**, 554–559

Webster, A. J. F. (1984) *Calf Husbandry, Health and Welfare*. Collins, London

Webster, A. J. F., Clarke, A. F., Madelin, T. M. and Wathes, C. M. (1987) Air hygiene in stables. 1. Effects of stable design, ventilation and management on the concentration of respirable dust. *Equine Veterinary Journal,* **19**, 448–453

Webster, A. J. F., Saville, C. and Welchman, D. de B. (1986) *Improved Husbandry Systems for Veal Calves.* University of Bristol Press, Bristol

Welchman, D. de B., Whelehan, O. P. and Webster, A. J. F. (1988) The haematology of veal calves reared in different husbandry systems and the assessment of iron deficiency. *Veterinary Record,* **123**, 505–510

White, D. G. (1986) Evaluation of a rapid, specific test for detecting colostral IgG$_1$ in the neonatal calf. *Veterinary Record,* **118**, 68–70

Chapter 14

New techniques in modelling cattle production systems

J. M. Bruce and P. J. Broadbent

Most decisions in our personal and professional lives are guided by models of various forms, although we are not often fully aware of this. Modelling is a most pervasive activity. Your response to an observed set of traffic conditions will be based on your view or model of what will or could happen next. Some accidents could be seen as the result of erroneous models or faulty interpretations. The Starch Equivalent method and the Metabolizable Energy method of assessing the nutritional requirements of livestock are both examples of models which also illustrate that different models can be acceptable as adequate for essentially similar purposes.

Models can be of a great variety of types. They can, for example, be pictures, verbal descriptions, tables, equations or computer programs. They can be empirical, mechanistic, static, dynamic, deterministic or probabilistic (France and Thornley, 1984). Indeed, it would be difficult to define all possible varieties of models. The role of modelling in communication has been emphasized by a number of authors (Zeigler, 1976; Burghes, Huntley and McDonald, 1982; Spedding, 1988). A model is a very compact way of packaging many related pieces of information which may be why the use of modelling has increased greatly in recent years.

In this chapter models of lactating suckler cows and growing cattle are used to illustrate the potential usefulness of modelling for simulating detailed aspects of livestock systems and for help in making strategic decisions. Breed type, feeding levels and schedules, and the use of shelter are examined. All the models used are mathematically deterministic and based on functional relationships. For cows the models consider only energy flows. The model for growing cattle considers protein, lipid, water and ash, and so calculates a body composition which can be more closely related to economic and nutritional values.

Notation

A	surface area of cow	m^2
C	net energy to conceptus	MJ/day
E	net energy balance	MJ/day
F	metabolizable energy (ME) in feed	MJ/day
I_a	external thermal resistance	$°Cm^2/W$
I_h	hair thermal resistance	$°Cm^2/W$
I_t	tissue thermal resistance	$°Cm^2/W$

$\bar{M}$	net energy in milk	MJ/day
Q_o	climatic energy demand	W/m^2
Q_e	evaporative heat loss	W/m^2
R	net radiation	W/m^2
r	rain	mm/h
T_a	air temperature	°C
T_b	deep body temperature	°C
V	wind speed	m/s
w	liveweight	kg
$\alpha, \beta, \gamma, \delta$	model parameters	

Climatic energy demand

Climatic energy demand is a thermal demand of the environment which arises when the ambient temperature is lower than that of the animal's deep body temperature. The concept applies particularly in the cold when the normal heat production by the animal may not match requirements for the maintenance of body temperature. In these situations the animal has to increase its heat production through an increase in metabolic rate and less energy is available for productive purposes. When the environmental conditions are less cold an animal's heat production is sufficient to maintain body temperature and it is said to be in the thermoneutral zone.

The work described here was carried out in conjunction with a project, at the North of Scotland College of Agriculture (NOSCA), which sought to evaluate the influence of winter environment on the performance and health of autumn-calving suckler cows and their calves. Within the NOSCA project 24 out-wintered cows were individually fed a nominal 80 MJ ME/day. The mean liveweight was 604 kg on 1 October 1978 at a mean of 23 days post-calving.

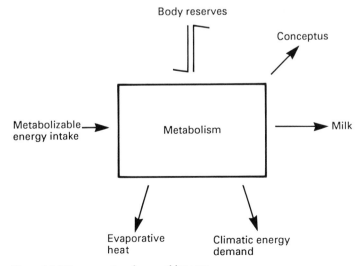

Figure 14.1 Energy system for a suckler cow

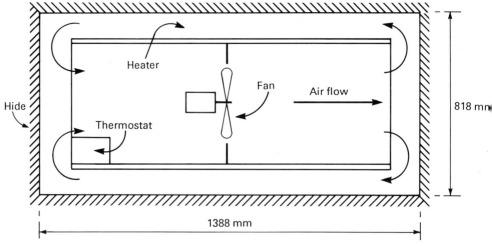

Figure 14.2 Thermal model of a suckler cow

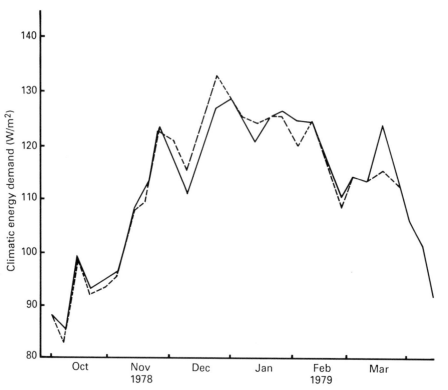

Figure 14.3 Climatic energy demand for winter 1978/79 (weekly mean). —— Estimated; ----- observed

Figure 14.1 shows the energy system for a suckler cow assumed in this work. Heat storage is ignored as it is a negligible factor in the long term. A thermal model of a suckler cow was located in the open near the out-wintering areas occupied by the experimental cows. The main features of the model are shown in Figure 14.2. The thermal insulation of a suckler cow is simulated using expanded polystyrene and a cured hide, complete with hair. An electronically controlled heater maintained the internal temperature at 39°C. The principle behind the model is this: if the internal temperature and the overall thermal insulation are approximately that of a real cow then the heat flow per unit surface area will be approximately that of a real cow. Every hour measurements were made of the climatic energy demand. Air temperature, wind speed, net radiation and precipitation were also automatically recorded hourly. The weekly mean climatic energy demand is shown in Figure 14.3 for the winter of 1978–79.

From Figure 14.1 we can write the energy balance equation:

$$E = F - A(Q_o + Q_e)/11.57 - M - C \qquad\qquad 1$$

The following assumptions are made:

$$A = 0.09\,w^{0.67}\,m^2, Q_e = 17\,W/m^2\ \text{(minimum)} \qquad\qquad 2$$

The energy balance was estimated for the cows without any correction for hair-coat depth, heat of warming or activity. The evaporative heat loss is a

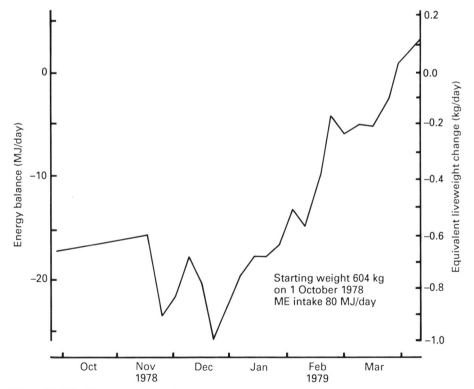

Figure 14.4 Weekly mean energy balance

minimum in cold conditions. The milk production was estimated from the mean growth of the calves and their mean feed intake other than milk. At a net energy for milk of 3 MJ/kg the estimated mean milk energy production is given by $9.0 - 1.02 \times 10^{-4} \, d^2$)kg/day) where d is measured in days from 1 October 1978.

Figure 14.4 shows the weekly mean net energy balance of the cows. Assuming liveweight loss to have a net energy content of 26.0 MJ/kg the maximum rate of liveweight loss was nearly 1 kg/day.

The cumulative liveweight change curve is shown in Figure 14.5. The agreement with the mean measurements for the 24 cows is sufficiently good to indicate the usefulness of the thermal model.

A mathematical model was developed in order to estimate the climatic energy demand on the thermal model using the four climatic variables measured. The following equation forms the basis of the model:

$$Q_0 = \frac{(T_b - T_a) - \alpha I_a R}{I_t + I_h + I_a} \qquad\qquad 3$$

where $I_a = 1/(5.3 + 7V^{0.6})$, $I_h = \beta(1 - \min(\gamma, \delta))$, $I_t = 0.203$, $T_b = 39$

The parameters were optimized to give the minimum error sum of squares between observed and predicted values of climatic energy demand.

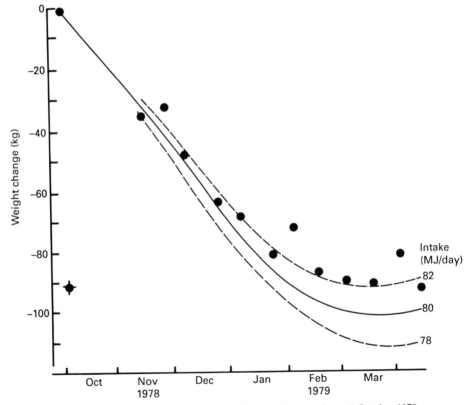

Figure 14.5 Cumulative liveweight change for a starting weight of 604 kg on 1 October 1978.
✦ Measured average for 24 cows

Table 14.1 Summary of errors* in climatic energy demand between model and data

	Number of values	Mean error (W/m^2)	Standard deviation (W/m^2)	Coefficient of variation (%)
Daily	201	−0.1 0.3	4.1 0.2	3.7 0.2
Weekly	29	−0.1 0.5	2.8 0.4	2.5 0.3

* Error = estimated − observed

Weekly values were calculated by averaging the seven daily mean values for climatic energy demand within calendric weeks

To estimate daily mean values of Q, daily mean climatic variables are used and the following parameters apply: $\alpha = 0.75$, $\beta = 0.068$, $\gamma = 1.5$, $\delta = 0.55$.

A summary of the errors is given in Table 14.1.

The estimated weekly mean values for climatic energy demand are shown in Figure 14.3. The agreement between observed and estimated is good although there is evidence that the errors are serially related.

The mean daily observed climatic energy demand was 112 W/m^2.

A thermal model of a suckler cow has been used to measure the climatic energy demand. Mathematical models have been developed to calculate the hourly and daily mean climatic energy demand using air temperature, wind speed, net radiation and precipitation. The mean liveweight change of 24 out-wintered suckler cows is predicted well using the measurements of climatic energy demand. It is concluded that physical and mathematical models can give good estimates of the climatic energy demand on suckler cows during winter in the north-east of Scotland.

The interaction of shelter and nutrition for suckler cows (Bruce, 1982)

The main purpose of a building is to provide shelter from the weather. It is very common to house cattle during winter but in the case of suckler cows housing is less frequently used. This stems mainly from the lack of finance as most agriculturalists intuitively believe that housing would be beneficial. At the present time, however, the benefits of housing suckler cows are not known. An assessment can be made, using a computer model, of the benefit of housing spring and autumn suckler cows at various levels of feeding. The benefits are measured in cow liveweight and metabolizable energy saved.

A second model was calibrated and placed in a naturally ventilated uninsulated, slatted cattle building occupied by suckler cows and calves. For simplicity a parabola was fitted to the mean daily values of climatic energy demand measured outdoors during the winter of 1978/79. This took the form:

$$Q_o = 91.8 + 0.62d - 0.00286d^2 \qquad\qquad 1$$

where d is the number of days from the start of October.

The measured climatic energy demand indoors showed a linear relationship with the climatic energy demand outdoors. The regression equation is:

$$Q_1 = 28.7 + 0.56Q_o \qquad\qquad 2$$

A building, as described above, apparently reduces the climatic energy demand by 20% in mid-winter and 13% in autumn and spring.

The spring-calving cows are assumed to be housed at the beginning of November at a maternal liveweight of 500 kg and turned out at the end of April. The liveweight, including conceptus, at housing is 512 kg. A 40 kg calf is assumed to be born during the first week of March.

The autumn-calving cows are assumed to be housed at the beginning of November at a liveweight of 500 kg and turned out at the end of April. It is assumed that a calf was born at the beginning of October and that the cow conceives 12 weeks from that time.

Milk production is assumed to be given by:

$$M = 8.0 - 0.05 N^2 \text{ kg/day} \hspace{3cm} \textbf{3}$$

where N is the week number from birth.

The milk is assumed to have a net energy of 3.0 MJ/kg and the energy value of liveweight change in the cow is assumed to be 26.0 MJ/kg.

A computer model was written which used the above equations. The recommendations of ARC (1980) were followed for energy requirements for maintenance and pregnancy, and for the various efficiencies of conversion of energy.

The model uses a time step of one week and flat rates of feeding are assumed. Protein is assumed non-limiting but may, in practice, highly influence both milk production and liveweight change.

Liveweight trajectories were calculated for feed levels of 50–100 MJ ME/day for spring- and autumn-calving cows both inside and outside. Figures 14.6 and 14.7 show some examples.

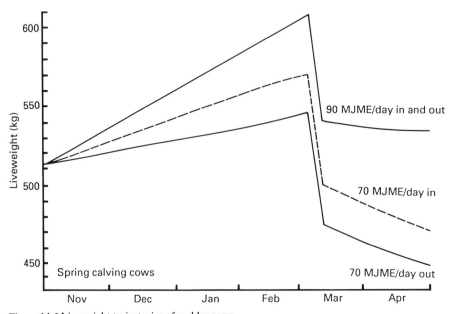

Figure 14.6 Liveweight trajectories of suckler cows

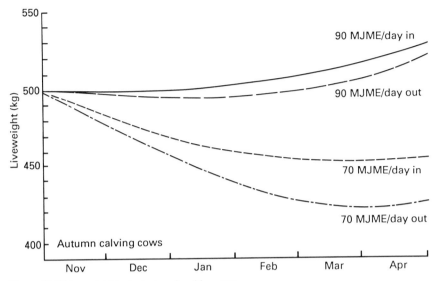

Figure 14.7 Liveweight trajectories of suckler cows

Figures 14.8 and 14.9 show the final liveweights at turnout related to feed level. From Figure 14.9, for example, it is seen that if an autumn calving cow were fed 75 MJ ME/day outside then it would lose 50 kg of liveweight from the beginning of November to the end of April. For the same liveweight change the housed cow need only be fed 69 MJ ME/day. If the housed cow were also fed 75 MJ ME/day then it would lose only 26 kg of liveweight. If a farmer decided that a liveweight loss of 60–70 kg was acceptable then housed cows could be fed about 8 MJ ME/day less

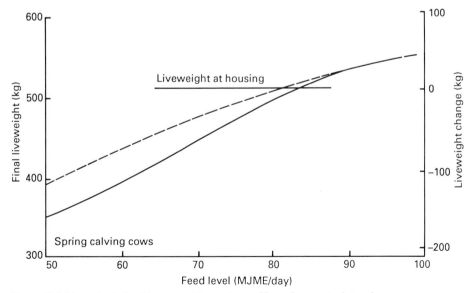

Figure 14.8 Liveweight of suckler cows at turnout. ----- Housed; —— out-wintered

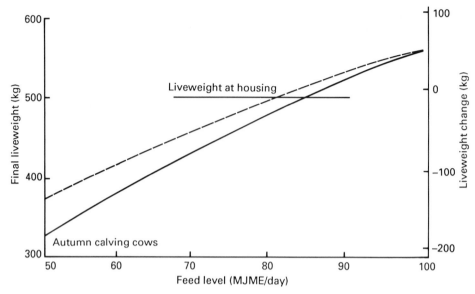

Figure 14.9 Liveweight of suckler cows at turnout. ----- Housed; —— out-wintered

than out-wintered cows. In this case the reduction in feed would be about the same for spring- or autumn-calving cows.

The benefits of housing, whether measured in cow liveweight or feed, are greater at low intakes and high liveweight loss. The indications are that we would expect to save a maximum of 8 MJ ME/day (0.7 kg/day of barley) by housing suckler cows in winter. It is likely to be much less than this.

The benefits of housing to suckler cows are suggested by the model to be small and by implication difficult to measure. The model results help towards more rational research and housing decisions.

Estimating liveweight changes of suckler cows when allowed shelter in a forest

There is a lack of research on the reduction of climatic energy demand on cattle when sheltering in a forest, although many studies have been made on the modification of individual climatic factors by forests. We can look at this problem using the mathematical model for the prediction of climatic energy demand on cattle using the climate variables of air temperature, windspeed, net radiation and rainfall. Daily mean values of the required climatic factors as measured at Craibstone, Aberdeen for 6 months starting on 1 October 1978 are used for predicting climatic energy demand values. Five shelter treatments are considered:

(1) fully exposed to the climate
(2) housed in a well-ventilated building
(3) 10 m into a pine forest
(4) 50 m into a pine forest
(5) 80 m into a pine forest.

Table 14.2 Climate modifiers

	Shelter treatment				
	1	*2*	*3*	*4*	*5*
Windspeed (m/s)	×1.00	×0.03	×0.55	×0.35	×0.10
Rainfall (mm/h)	×1.00	×0.00	×0.80	×0.80	×0.80
Air temperature (°C)	+0.00	+1.00	+0.00	+0.50	+1.00
Net radiation (W/m^2)	×1.00	×0.05	×0.10	×0.10	×0.10

The climate modifiers (Table 14.2) were estimated from the results of research on forest climatology. An energy balance was calculated at the end of each week for each treatment using the liveweight at the start of the week.

If the liveweight changes obtained in the building (i.e. lowest climatic energy demand) were desired in the other shelter treatments then extra feed would be required.

Table 14.3 Diets fed to cattle

MJ ME/day	kg/day		Cost
			(£/day)
	Silage	*Barley*	
70	25.5	1.30	0.45
80	29.1	1.46	0.51
90	32.7	1.70	0.58

Dietary constituents were:
  Silage 22% DM, 10 MJ ME/kg DM, £12/tonne
  Barley 85% DM, 13 MJ ME/kg DM, £110/tonne.

Table 14.4 Number of extra days feed required to maintain the same liveweight change as that in a building over a 28 week winter, and the cost (£/cow) this represents for the whole winter (in parentheses)

Feed intake (MJ ME/day)	Shelter treatment							
	1		*3*		*4*		*5*	
70	29.4	(13.23)	22.5	(10.13)	17.5	(7.88)	7.4	(3.33)
80	13.0	(6.63)	7.6	(3.89)	5.3	(2.70)	0.0	(0.00)
90	5.0	(2.90)	1.9	(1.10)	0.7	(0.41)	0.0	(0.00)

The number of days extra feed required could then be estimated and the cost per cow calculated from Table 14.3. The results (Table 14.4) show that these estimated savings are very low and it is likely that eliminating the errors in estimating the various parts of the model would show little change in these results.

The importance of shelter for suckled calves

The thermal model previously described is too large and expensive for use in shelter studies where a number of models are required for comparison under the same climatic conditions. Four smaller versions were made to a scale of about 1:0.4. A synthetic coat was used because of difficulty in obtaining sufficiently similar 'real' coat samples. The synthetic material chosen had similar thermal characteristics to a 'real' coat when tested under various climatic conditions.

The shelters used in this experiment represented four basic stages of climatic modification by simple shelters. These were:

(1) no shelter
(2) four walls (solid, no roof)
(3) a roof and no walls
(4) roof and walls, with ventilation openings simulating a well-ventilated cattle building.

These structures were of plywood board construction and are about 2.5 m square and 1.5 m high. One of the small thermal models was placed in each of the shelter treatments. Heat loss and temperature measurements were automatically recorded onto cassette tape every hour. Data were also collected from the large (unsheltered) thermal model sited nearby (39 m) for comparison with the unsheltered small model.

Results

The ratio of climatic energy demand (CED) values for each model to the CED value for the unsheltered small model was calculated hourly. Table 14.5 shows the results for 14 weeks' data starting on 13 December 1981. Analysis shows a highly significant difference between the shelter treatments for the four small models ($P<0.001$) and a non-significant difference between the unsheltered small and large models ($P>0.05$). The differences between shelter types, although small, were consistently measured with the models. The reduction of CED by 20% as measured by the small model in the simulated cattle building (walls and roof), is similar to the results from the comparison of the two large models over the previous 2 years where one was placed outside and the other was placed in a functioning cattle building. This demonstrates the similarity of the large and small models and the simulated and real buildings.

Table 14.5 Mean ratio of climatic energy demand (CED) values for each model to the CED value for the unsheltered small model

Models and degrees of shelter		Mean	Standard error
Large model (unsheltered)		0.998	0.001
Small models	unsheltered	1.000	–
	roof only	0.983	0.001
	walls only	0.942	0.001
	walls and roof	0.801	0.001

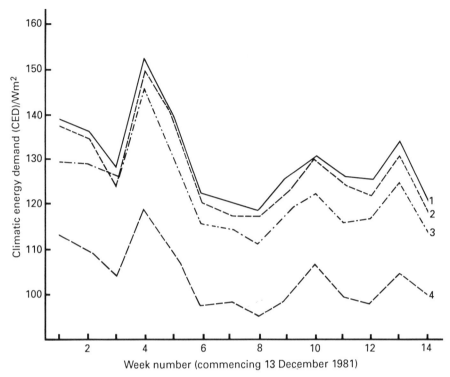

Figure 14.10 Weekly mean climatic energy demand (CED) values measured for each shelter. 1, Unsheltered small model; 2, roof only small model; 3, walls only small model; 4, roof and walls small model

Figure 14.10 gives weekly mean CED values for the four small models and demonstrates the shelter effects found. The similarity of the large and small unsheltered models previously found (Table 14.5) makes it likely that the mathematical model can be applied to these smaller models.

The overall aim of this work was to estimate CED for cattle which are protected to some degree by a form of shelter. The following is an example of how these predictions were applied to a practical situation where a group of suckler cows and calves were kept outside all winter at Aberdeen. The calves had access to a creep area with solid walls but no roof. This work investigated the benefits obtained from this type of shelter.

Climatic energy demand values for the calves were estimated from the mathematical model. The climatic variables used were weekly mean values measured at the experimental site between October 1978 and April 1979 (29 weeks). Surface area and tissue insulation values were altered for application of the model to calves.

When the total heat loss from the calves (sensible CED and evaporative) is in excess of thermoneutral heat production then metabolizable energy intake is redirected from liveweight gain to heat production for the maintenance of homeothermy. Thermoneutral heat production can be calculated from estimates of

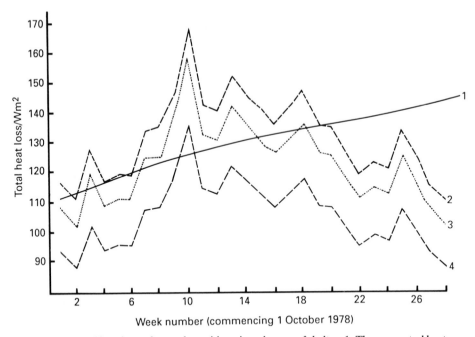

Figure 14.11 Total heat losses from calves with various degrees of shelter. 1, Thermoneutral heat production; 2, no reduction of climatic energy demand (CED); 3, 0.94 × unsheltered CED; 4, 0.8 × unsheltered CED

liveweight (40 kg on 1 October), liveweight gain (0.9 kg/day) and the dietary constituents of the calves using ARC (1980) recommendations.

The results for total calf heat losses and thermoneutral heat production are shown in Figure 14.11. Total heat losses from the calves are also calculated when estimated CED values are multiplied by 0.94 and 0.80 simulating a calf creep and a cattle building respectively (Table 14.5).

The impact of heat losses in excess of thermoneutral heat production can be expressed in terms of the reduction of liveweight gain through the redirection of feed metabolizable energy towards heat production. Table 14.6 gives estimated cumulative liveweight deficits for each shelter treatment for the 29-week winter.

Table 14.6 Cumulative liveweight deficit for each shelter treatment for the whole winter (29 weeks)

Shelter	Estimated cumulative liveweight deficit (kg)
No shelter	40
0.94 × CED	20
0.80 × CED	2

CED, climatic energy demand.

The results show that a small reduction in heat loss by a simple walled shelter can halve the reduction of liveweight gain resulting from climatic effects. This demonstrates the importance of evaluating these relatively small reductions in heat loss in terms of the benefits to the animal. There is room for improvement of various aspects of the analysis but the general usefulness of modelling techniques has been shown when applied to practical situations.

NOSCOW – A model of the energy system of lactating and pregnant cows

The previous calculations for cows always assumed that a reasonable estimate of milk production could be made. The NOSCOW model (Bruce, Broadbent and Topps, 1984) does not require this assumption. This model was designed to answer the question: what happens to milk production and liveweight when a lactating, pregnant cow eats a given amount of metabolizable energy? The model includes a Gompertz growth model and an equation for a lactation curve given by Wood (1980). The model is firmly based on empirical recommendations (ARC, 1980) and, importantly, includes genetic potentials for growth and lactation and a mechanism for balancing these potentials. This makes possible the simultaneous prediction of milk production and liveweight change.

It is outside the scope of this chapter to present the details of NOSCOW. The emphasis is not on the model but on the results and the usefulness for practical consideration.

The following examples illustrate the application of modelling to suckler cows all use NOSCOW.

The effect of winter food level on milk production and liveweight of autumn-calving suckler cows

Four winter levels of uncorrected metabolizable energy intake are considered. These are 70, 85, 100 and 115 MJ/day. Calving is assumed to be at the beginning of October at Aberdeen. The winter food levels are maintained for 32 weeks from calving, followed by 10 weeks at 120 MJ/day and a further 10 weeks at 80 MJ/day to simulate grazing. Conception is assumed to take place at 12 weeks after calving and the length of the lactation period is 42 weeks.

Figure 14.12 shows the model predictions of milk energy production. The results show a marked effect of winter nutrition on the levels of milk energy production. For the 100 and 115 MJ/day intake levels the calf is limiting milk energy production in the early part of lactation. The effect of the increase in intake to 120 MJ/day, following turn-out to grazing at week 32, is to cause an increase in milk energy production that is large for lower-fed cows. The early grazing is clearly important to these cows. For the 70 MJ/day cow, 30% of the total milk energy production takes place at grazing. The total milk produced is given in Table 14.7. To complete the agricultural picture, the milk production would need to be considered in the context of calf productivity and food costs using different dietary sources.

The liveweight trajectories are shown in Figure 14.12 and some points are summarized in Table 14.7. The minimum liveweight and the final post-calving liveweight increase with winter food level. If a management objective were to return the cow to its initial post-calving liveweight then the 85 MJ/day winter level is

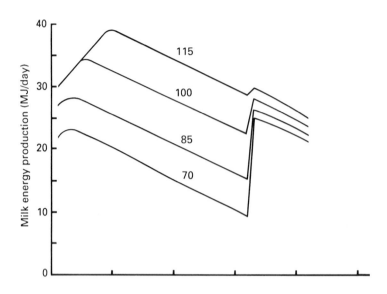

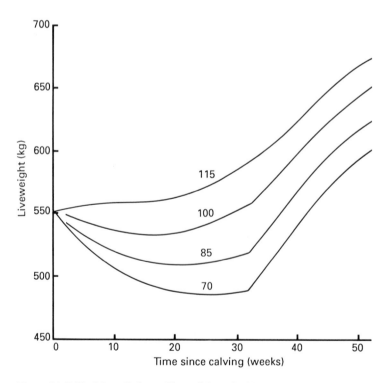

Figure 14.12 Model predictions of liveweight and milk energy production for suckler cows. The curves are marked with the winter (weeks 0–32) metabolizable energy intake (MJ/day)

Table 14.7 Model results for suckler cows weighing 550 kg at initial calving with different winter (weeks 0–32) metabolizable energy intakes

Winter ME intake (MJ/day)	70	85	100	115
Total annual ME intake (MJ)	29 680	33 040	36 400	39 760
Minimum liveweight (kg)	485	509	534	550
Final post-calving liveweight (kg)	529	550	576	601
Maximum rate of maternal liveweight loss (kg/day)	−0.66	−0.48	−0.40*	−0.44*
Total milk energy produced (MJ)	5 370	6 710	8 283	9 529

* During the last week of pregnancy.

suggested, by the model, to be appropriate for unexposed cows. The greatest rate of maternal liveweight loss is experienced by the lowest-fed cow. It is a matter of judgement whether this will prevent the establishment of a viable pregnancy at week 12. For the two highest food levels the greatest rates of maternal liveweight loss are during the last week of pregnancy: 80 MJ/day is insufficient for maintenance and pregnancy during the late stages of pregnancy.

Further details of this example may be seen in Bruce, Broadbent and Topps (1984).

The nutrition of twin-bearing cows during pregnancy and lactation (Topps, Broadbent and Bruce, 1985)

Twin-bearing and twin-suckling suckler cows of 550 kg liveweight with a potential mature liveweight of 700 kg and a potential peak milk energy yield of 0.75 MJ/kg$^{0.75}$ are assumed to be housed during winter. Each calf weighed 35 kg at birth on 1 October and had a potential daily gain of 0.85 kg.

The cows received one of five energy intakes, 90, 100, 110, 120 or 130 MJ ME/day, given as diets with an energy content of 8.5, 9.0, 9.5, 10.0 and 10.5 MJ ME/kg DM respectively from calving until the calves were weaned at 24 or 32 weeks. For 8 weeks following weaning at week 24 energy intake was 59 MJ ME/day from a diet of 7.0 MJ ME/kg DM, which represented the feeding to appetite of a poor quality diet for the last part of winter. After week 32 energy intake was 120 MJ ME/day provided by a diet of 11.0 MJ ME/kg DM for a period of 10 weeks, a dietary regimen which was considered to be similar to that provided by early summer grazing. For the final 10 weeks, the cows received one of three energy intakes, 80, 90 or 100 MJ ME/day each from diets of 11.0 MJ ME/kg DM which was considered to be similar to that provided by late summer grazing differing in amounts available to the animal due to differences in grazing pressure. Thus 15 combinations of winter and late summer nutrition were examined for cows whose twin calves were weaned at either 24 or 32 weeks of age.

The acceptability of the calculated performance was judged by considering milk yields, minimum liveweight of the cow during the 52 weeks and at conception (12 weeks after calving), total ME intake and the post-calving liveweight of the cow which was required to be 550 ± 10 kg.

Table 14.8 shows the milk energy yield of the cows for a lactation of either 24 or 32 weeks. For a lactation of 24 and 32 weeks, yields of milk energy ranged from 5730 to 8500 and from 7040 to 10780 MJ respectively. Twin calves growing at 0.85 kg/day would require approximately 2 × 3500 or 4500 MJ for a suckling period

Table 14.8 Milk energy produced in lactation of either 24 or 32 weeks (MJ)

Winter energy intake (MJ ME/day)	24 weeks	32 weeks
90	5730	7 040
100	6400	7 940
110	7090	8 870
120	7790	9 820
130	8500	10 780

Table 14.9 Total ME intake (GJ) over 52 weeks of cows with a lactation of either 24 or 32 weeks

Winter energy intake (MJ ME/day)	24 weeks			32 weeks		
	Late summer intake (MJ ME/day)			Late summer intake (MJ ME/day)		
	80	90	100	80	90	100
90	32.4	33.1	33.8	34.1	34.8	35.5
100	34.1	34.8	35.5	36.4	37.1	37.8
110	35.8	36.5	37.2	38.7	39.4	40.1
120	37.5	38.2	38.9	40.9	41.6	42.3
130	39.1	39.8	40.5	43.1	43.8	44.5

of 24 or 32 weeks respectively. Such yields are provided by a winter energy intake of 110 MJ ME/day. However, with the judicious use of solid food, calf growth rates of 0.85 kg/day may be maintained on lactation yields of 3200 MJ for 24 weeks or 4000 MJ for 32 weeks, so a winter energy intake of 100 MJ ME/day may be acceptable.

Table 14.9 shows the annual ME intake of the cows for a lactation of either 24 or 32 weeks. For the shorter and longer lactations the difference between the lowest and highest intake was 8.1 and 10.4 GJ respectively. It is important that the least amount of food is used to produce enough milk for two good calves so levels of 110 MJ ME/day during winter and of 80 MJ ME/day in later summer are the most appropriate. However, a winter energy intake of 100 MJ ME/day with the same late summer nutrition may also be acceptable.

Tables 14.10 and 14.11 give the weight loss of the cows from calving to 12 weeks post-partum and the minimum weight of the cows respectively. If the weight loss in the 12 weeks following calving is greater than 0.15 of the post-calving liveweight the chances of re-conception are considered to be appreciably reduced. In these cases energy intakes would not be acceptable, but none of the 15 combinations of nutritional treatments was in this category. Similarly, if the weight loss at any time from calving exceeded 0.30 of the post-calving liveweight, it may have a serious effect on both the viability and fertility of the cow and these treatments would be unacceptable. None of the 15 combinations led to such large losses of liveweight.

Table 14.10 Weight loss of cows (kg) from calving to 12 weeks post-partum with a lactation of either 24 or 32 weeks

Winter energy intake (MJ ME/day)	Weight loss (kg)
90	68
100	58
110	49
120	39
130	29

Table 14.11 Minimum weight (kg) in the annual cycle (week of its occurrence from parturition) of cows with a post-calving liveweight of 550 kg and a lactation of either 24 or 32 weeks

Winter energy intake (MJ ME/day)	24 weeks	32 weeks
90	453 (24)	449 (32)
100	469 (24)	468 (28)
110	485 (24)	485 (25)
120	502 (22)	502 (22)
130	516 (19)	516 (19)

Table 14.12 Subsequent post-calving liveweight of cows (kg) weaned after 24 or 32 weeks of lactation

Winter energy intake (MJ ME/day)	24 weeks			32 weeks		
	Late summer intake (MJ ME/day)			Late summer intake (MJ ME/day)		
	80	90	100	80	90	100
90	525	539	553	518	532	545
100	536	550	564	533	547	560
110	548	562	575	547	563	575
120	559	573	587	562	577	590
130	571	585	598	577	592	605

Table 14.12 shows the post-calving liveweight of the cows following a lactation of either 24 or 32 weeks. For each lactation the weights differed by as much as 70 kg or over which is considered to be the equivalent of one unit of condition score. However, an essential criterion is that the cow returns to her post-calving liveweight of 550 ± 10 kg. As a result, and bearing in mind the milk yield required, a winter energy intake of 110 MJ ME/day together with a late summer intake of

80 MJ ME/day is the most acceptable of the 15 combinations studied. An alternative, if calf growth rates can be sustained by the use of solid feeds, is 100 MJ ME/day during winter and 90 MJ ME/day during the summer.

Trials carried out in Aberdeen over the first 23 weeks of lactation with Hereford × Friesian cows calving in November and suckling a natural and a foster calf have shown that a winter energy intake of 110 MJ ME/day is an acceptable level. The cows invariably calved in good condition but both their weight loss at 12 weeks' post-partum and the maximum loss were akin to that predicted by the model. Milk yields of the cows, measured by the weigh-suckle-weigh technique, were equivalent to or better than those predicted.

The simulated performance of suckler cows having various combinations of potential size and milk production, given different levels of winter energy intake (Broadbent, Bruce and Topps, 1985)

The NOSCOW model was used to calculate the performance of suckling cows of various potential mature liveweights and potential peak milk yields given different levels of winter feed. The results were calculated for a factorial arrangement of five potential mature liveweights (kg), five potential peak milk energy yields (MJ/kg$^{0.75}$) and three levels of maximum winter energy intake (MJ ME/day) giving 75 combinations of genotype and winter energy intake.

The three levels of maximum winter energy intake and their associated dietary energy concentration (MJ ME/kg DM) were 70, 8.5; 85, 9.5; and 100, 10.5. These were applied for 32 weeks from calving on 1 October. Outdoor climatic conditions appropriate to Aberdeen, Scotland were simulated and weaning occurred after 32 weeks of lactation. The 32-week winter period was followed in all cases by two consecutive periods of 10 weeks each with minimum energy intake of 120 and 80 MJ ME/day at 11.0 MJ ME/kg DM.

The five potential mature liveweights used were 600, 675, 750, 825 and 900 kg and the five potential peak milk yields were 0.20, 0.45, 0.70, 0.95 and 1.20 MJ/kg$^{0.75}$.

Each combination of genotype and winter energy intake was simulated from an initial liveweight of 0.40 potential mature liveweight, when sexual maturity was

Table 14.13 Summary of circumstances simulated

Circumstance	Units	Level(s) simulated
Potential mature liveweight	kg	600, 675, 750, 825, 900
Potential peak milk yield	MJ/kg$^{0.75}$	0.20, 0.45, 0.70, 0.95, 1.20
Maximum energy intake (dietary energy concentration)		
32 weeks' winter (lactation)	MJ ME (MJ ME/kg DM)	70, (8.5); 85, (9.5); 100, (10.5)
10 weeks' summer	MJ ME (MJ ME/kg DM)	120 (11.0)
10 weeks' summer	MJ ME (MJ ME/kg DM)	80 (11.0)
Calving date	–	1 October
Lactation	weeks	32
Conception	week	12
Climate	–	Aberdeen, Scotland
Initial liveweight	kg	0.40 potential mature liveweight
Calf birthweight	kg	0.05 potential mature liveweight
Calf growth rate	kg/day	0.0011 potential mature liveweight

Table 14.14 Stable post-calving liveweight (kg) and milk energy produced in subsequent lactation (MJ/32 weeks)

Energy intake over 32 weeks of winter

Potential peak milk energy yield (MJ/kg$^{0.75}$)

(a) Stable post-calving liveweight (kg)

Potential mature liveweight (kg)	70 MJ ME/day					85 MJ ME/day					100 MJ ME/day				
	0.20	0.45	0.70	0.95	1.20	0.20	0.45	0.70	0.95	1.20	0.20	0.45	0.70	0.95	1.20
600	600	542	490	442	350	633	577	537	502	477	671	619	583	568	558
675	624	562	503	457	399	670	603	549	507	475	710	647	599	565	543
750	641	581	518	465	427	700	625	563	512	480	743	668	611	564	530
825	639	599	529	478	434	726	645	572	528	482	772	688	618	570	536
900	642	616	540	485	433	742	663	582	531	480	798	709	632	586	533

(b) Milk produced at stable post-calving liveweight (MJ/32 weeks)

Potential mature liveweight (kg)	70 MJ ME/day					85 MJ ME/day					100 MJ ME/day				
	0.20	0.45	0.70	0.95	1.20	0.20	0.45	0.70	0.95	1.20	0.20	0.45	0.70	0.95	1.20
600	1140	3160	4650	5640	5360	3030	4890	6130	6890	7320	4580	6390	7270	7670	7810
675	500	2570	4330	5540	6290	2300	4430	5920	6920	7560	4230	6090	7300	8010	8390
750	190	2030	3970	5390	6390	1590	3900	5680	6870	7710	3490	5690	7190	8150	8740
825	50	1520	3570	5230	6360	900	3370	5370	6810	7760	2780	5230	6970	8180	8980
900	0	1040	3200	5010	6260	450	2870	5020	6660	7750	2110	4730	6740	8180	9100

Table 14.15 Combinations of genetic potential mature liveweight (kg) and peak milk yield (MJ/kg$^{0.75}$) which produce acceptable levels of milk (MJ) at each level of winter energy intake (MJ ME/day over 32 weeks)

	Energy intake over 32 weeks of winter														
	70 MJ ME/day					85 MJ ME/day					100 MJ ME/day				
Potential peak milk energy yield (MJ/kg$^{0.75}$)	0.20	0.45	0.70	0.95	1.20	0.20	0.45	0.70	0.95	1.20	0.20	0.45	0.70	0.95	1.20
Potential mature liveweight (kg) 600			*				*				*				
675				*				*				*			
750				(*)				*				*			
825					(*)				*				*		
900									*				*		

Note: Parentheses indicate that the cow may not conceive at week 12 of lactation or may not be viable due to the degree of liveweight loss.

assumed to have been achieved. In the initial cycle the cow was assumed to conceive at week 12 of the simulation and in each cycle thereafter conception occurred at week 12 of lactation. The simulation was continued until the increase in post-calving liveweight from one parturition to the next was less than $0.01 \times$ the current post-calving liveweight. The penultimate post-calving liveweight simulated was regarded as the stable liveweight appropriate to that genotype-nutritional combination. In most cases this stable situation was achieved at the fourth calving. Higher energy intakes than those designated were not allowed. Within this constraint dry matter (kg DM/day) and, hence, energy (MJ ME/day) intake was restricted to $0.0225 \times$ maternal liveweight at the start of the winter period of 32 weeks, and $0.0250 \times$ maternal liveweight at the start of each 10-week summer period unless a higher maternal liveweight had been achieved previously at the beginning of any period, in which case intake was related to this higher liveweight. Table 14.13 summarizes the circumstances simulated.

Table 14.14 shows the stable post-calving liveweight (kg) and the milk energy yield (MJ) achieved in the subsequent 32 weeks of lactation. Those combinations of genotype and winter energy intake which provide an acceptable, but not excessive, level of milk yield have been indicated in Table 14.15.

For cows fed 70 MJ ME/day during winter having a potential mature liveweight of 750 kg and potential peak milk energy yields of 0.95 or $1.20 \text{MJ/kg}^{0.75}$ the maternal liveweight at week 12 of lactation was marginally less than 0.85 of the most recent post-calving liveweight which suggests that conception may not be achieved at that point. This problem would be obviated in these two instances by a small increase in winter energy intake. Alternatively, housing the cows during winter would reduce the body energy mobilized sufficiently to enable fertility to be maintained. At the same feeding level a cow with genetic potentials of 825 kg, $1.2 \text{MJ/kg}^{0.75}$ was also less than 0.85 of its most recent post-calving liveweight at week 12 of lactation, and its minimum maternal liveweight within a parity was less than 0.70 of its highest post-calving liveweight giving question to both its fertility and viability. The results show (Table 14.16) that as the winter energy available decreases milk potential must increase to maintain a suitable level of output at any particular potential mature liveweight. But at moderate to high potential liveweight this may be accompanied with fertility or viability problems. Although the combination of low genetic potentials (600 kg, $0.2 \text{MJ/kg}^{0.75}$) produces suitable results, at a high winter energy intake (100 MJ ME/day) the cow will tend to be obese and have a high or very high body condition score (Table 14.16).

Table 14.16 Winter feed levels (MJ ME/day over 32 weeks) which allow acceptable milk production (MJ) for various combinations of potential mature liveweight (kg) and potential peak milk yield (MJ/kg$^{0.75}$)

		Potential peak milk yield (MJ/kg$^{0.75}$)				
		0.20	0.45	0.70	0.95	1.20
Potential mature liveweight (kg)	600	100	85	70		
	675		100	85	70	
	750		100	85	(70)	(70)
	825			100	85	(70)
	900			100	85	

Note: parentheses indicate that the cow may not conceive at week 12 of lactation or may not be viable due to the degree of liveweight loss.

It is concluded that suitable genetic combinations of potential mature liveweight (kg) and potential peak milk energy yield (MJ/kg$^{0.75}$) for each level of winter energy intake (MJ ME/day over 32 weeks) are as follows:

70 MJ ME/day: 600, 0.70 and 675, 0.95
85 MJ ME/day: 600, 0.45; 675, 0.70; 750, 0.70; 825, 0.95 and 900, 0.95
100 MJ ME/day: 600, 0.20; 675, 0.45; 750, 0.45; 825, 0.70 and 900, 0.70.

(Note: For a Hereford × British Friesian potential mature liveweight is around 750 kg and potential peak milk yield around 0.70 MJ/kg$^{0.75}$.)

Growth of beef animals

Predicting the growth response of beef animals to feeds containing different constituents is a difficult but necessary calculation. ARC (1980) makes this possible by using equations derived from empirical observations of growth and the efficiencies related primarily to energy. The important ideas of rumen degradable protein (RDP) and undegradable protein (UDP) have emerged and allow a better estimation of protein requirements. Ingvartsen (1988) and Korver (1988) use essentially empirical or statistical models to estimate growth in bulls. Emmans and Oldham (1988) suggest a more theoretical approach to growth production based on Gompertzian potential growth. Loewert et al. (1983) use empiricism within a theoretical framework. The model used for the purposes of this chapter is most similar to the scheme suggested by Emmans and Oldham (1988) although not identical. A Gompertzian growth potential for protein mass is assumed with proportional water and ash. The feed is described by metabolizable energy, RDP and UDP. The model is a time series with steps of one day and the weight change is the sum of protein, water, ash and fat. Although the model is set up to use partial coefficients of utilization of metabolizable energy for protein and fat these are both set to a single value equal to the efficiency of utilization for whole body growth given by ARC (1980) as a function of the metabolic potential of the diet.

Table 14.17 shows examples taken from Wilkinson (1985). The mature protein masses assumed are 100, 104 and 113 kg for Hereford × Friesian, Friesian, and Charolais × Friesian respectively. The very good accuracy with which the current, more complex, model agrees with the values given by Wilkinson (1985) for final weight is not the most important consideration. What is of most importance is that a growth model based on the potential for protein deposition, with fat growth as a consequence of the energy balance, can produce results similar to current practice but which inherently produces more information about the composition of the animal at slaughter. The protein to fat ratio estimated by the model is included in Table 14.17. The range from 0.57 to 0.65 indicates that there are likely to be very important differences in the carcasses. A reliable model which allowed protein to fat ratios to be estimated, even relatively if not absolutely, would be extremely useful for determining agricultural strategies. Figure 14.13 illustrates for one case the growth of the body components.

Table 14.17 Alternative feeding strategies with different genotypes of cattle

Wilkinson (1985) model	*Hereford × Friesian*					*Charolais × Friesian*			*Friesian*		
Barley (kg/day)	1.5	2.0	2.4	2.5	3.0	2.5	3.0	3.5	2.2	2.6	3.0
Silage (kg DM/day)	5.8	5.0	4.8	5.3	4.8	5.3	4.8	4.3	5.0	4.5	4.3
Period (days)	210	215	155	200	160	210	185	160	285	220	165
Start weight (kg)	315	325	311	315	315	360	360	360	326	324	327
Final weight (kg)	490	475	450	475	460	530	525	520	525	500	475
Current model											
Prediction of final weight (kg)	478	496	445	488	463	542	529	514	533	499	472
Difference (kg)	−12	−21	−5	+13	+3	+12	+4	−6	+8	−1	−3
Protein/fat ratio from current model	0.65	0.64	0.65	0.57	0.58	0.59	0.58	0.57	0.62	0.63	0.62

After Wilkinson, 1985.

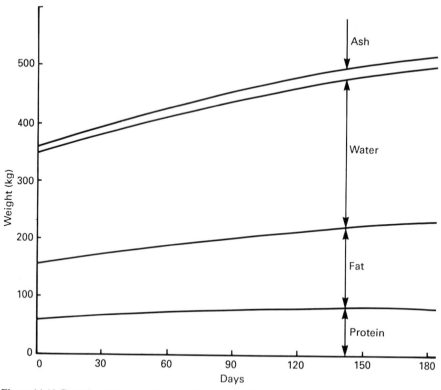

Figure 14.13 Growth and composition for Charolais × Hereford cattle fed 3 kg barley and 4.8 kg DM of silage

Conclusions

Modelling is pervasive and no more so than in animal production. Cases have been presented which illustrate how models can be used to integrate the effects of climate, shelter, feed and genotype on growth, weight loss, lactation and body composition. The types of models described here will be in common use in agriculture in the near future. They are already being used to guide research into the crucial areas. Even in the difficult areas of lactation and growth the models show good accuracy and would appear to be adequate for strategic decisions.

For the future we must look for improvement but it is already seen that further steps are required. For example, the relationships between body composition and carcass quality, economic value, nutritional value and palatability need to be modelled. In essence what is required is a model which converts animal feed variables into variables describing the value of the resultant food for humans: a feed-to-food model.

References

Agricultural Research Council (ARC) (1980) *The Nutrient Requirements of Ruminant Livestock.* Commonwealth Agricultural Bureaux, Slough

Broadbent, P. J., Bruce, J. M. and Topps, J. H. (1985) The simulated performance of suckler cows having various combinations of potential size and milk production, given different levels of winter energy intake. *Animal Production,* **40**, 554–555 (abstract)

Bruce, J. M. (1982) The effect of feed level and environment on suckler cows. *Animal Production,* **34**, 393–394 (abstract)

Bruce, J. M., Broadbent, P. J. and Topps, J. H. (1984) A model of the energy system of lactating and pregnant cows. *Animal Production,* **38**, 351–362

Burghes, D. N., Huntley, I. and McDonald, J. (1982) *Applying Mathematics.* Ellis Horwood Limited, Chichester

Emmans, G. C. and Oldham, J. D. (1988) In *Modelling of Livestock Production Systems,* (ed. S. Korver and J. A. M. Van Arendonk). Kluwer Academic Publications, Dordrecht, pp. 22–31

France, J. and Thornley, J. H. M. (1984) *Mathematical Models in Agriculture.* Butterworths, London

Ingvartsen, K. L. (1988) In *Modelling of Livestock Production Systems,* (ed. S. Korver and J. A. M. Van Arendonk). Kluwer Academic Publications, Dordrecht, pp. 22–31

Korver, S. (1988) In *Modelling of Livestock Production Systems,* (ed. S. Korver and J. A. M. Van Arendonk). Kluwer Academic Publications, Dordrecht, pp. 32–38

Loewert, O. J., Smith, E. M., Taul, K. L., Turner, L. W. *et al.* (1983) A body composition model for predicting beef animal growth. *Agricultural Systems,* **10**, 245–256

Spedding, C. R. W. (1988) In *Modelling of Livestock Production Systems,* (ed. S. Korver and J. A. M. Van Arendonk). Kluwer Academic Publishers, Dordrecht, pp. 3–9

Topps, J. H., Broadbent, P. J. and Bruce, J. M. (1985) The nutrition of twin-bearing cows during pregnancy and lactation. *Animal Production,* **40**, 528 (abstract)

Wilkinson, J. M. (1985) *Beef Production from Silage and Other Conserved Products.* Longman, London, pp. 24–25

Wood, P. D. P. (1980) Breed variations in the shape of the lactation curve of cattle and their implications for efficiency. *Animal Production,* **31**, 133–141

Zeigler, B. P. (1976) *Theory of Modelling and Simulation.* John Wiley and Sons, New York

Chapter 15

The incorporation of new techniques into cattle production systems in developed countries

J. D. Leaver

In developed countries the rate of incorporation of new techniques into farming systems is influenced not only by their rate of appearance (which depends on the amount of research and development activity), but also by the suitability of the techniques for individual farm circumstances.

Surpluses of milk and dairy products, and of beef in many developed countries are likely to be major influences on which techniques are incorporated in future. The stabilizing of national populations and the trend towards a declining consumption of some cattle products, combined with increasing efficiencies of cattle production systems, will place greater pressures on farmers. The objective of raising the productivity of cattle and the output of farms is ceasing to be the most viable way forward.

In addition to financial constraints, farmers are faced with concerns being expressed over other issues including pollution from cattle units, cattle welfare in intensively managed systems, the biological manipulation of cattle performance and the relationship between the consumption of cattle products and human health. Thus, although research is generating new ideas potentially capable of being incorporated into cattle production systems, these will in future be more stringently examined for economic, environmental and social considerations than in the past.

Transfer of technology

The dynamic nature of agricultural industries in developed countries, allows in general a rapid transfer of technology between research and practice. The phases of transfer are outlined in Figure 15.1.

The duration and intensity of the development phase varies according to the individual technique. Development involves not only validating the technique under farming conditions, but also complying with any legislative regulations which may be necessary before incorporation. For techniques involving manufactured products, the necessary marketing structure will also have to be developed. Many techniques fail to progress further than this phase due to failure to perform adequately under farm conditions.

Publicity for new techniques normally occurs during the development phase. This may be from the research team speaking at conferences, and via written articles and scientific papers. Progressive farmers and extension workers are therefore informed quickly about new developments. Techniques and products

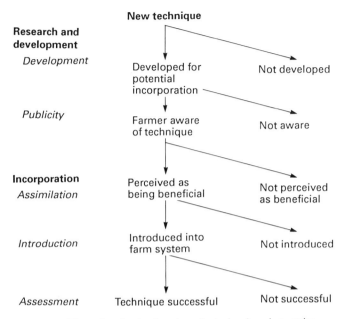

Figure 15.1 Hierarchy of technology transfer in developed countries

marketed by companies from the ancillary industries to agriculture also have strong advertising backing. Government and company extension agencies supplement communication about new techniques through farm walks, discussion groups, meetings and leaflets. The farmer in developed countries is therefore in a favourable position to hear at an early stage about new technologies. Nevertheless, the smaller farmer who has less time to spend away from the farm, is less likely to hear about, and be involved in discussions concerning new techniques. Farm scale is therefore a most important factor in technology transfer in both developing and developed agriculture. In addition to being less likely to hear about new techniques, the smaller farmer also has less flexibility to incorporate change due not only to financial constraints but also to the close integration of components within the system. A change in one component under such conditions causes interactions with other components and, as a result, the outcome may be less predictable than on larger farms.

The assimilation phase for the farmer who has heard of a new technique is the period during which he contemplates whether it should be incorporated. In developed countries there are many extension agencies who will provide the technical and financial advice to assist the decision-making process. Also, an increasing proportion of farmers have received higher educational training in agriculture and are more likely to have the ability to decide if, when and how a technology should be incorporated onto the farm, and what the implications are likely to be.

Following the assimilation process, if the technique is perceived as having potential benefit, it may be introduced into the farming system. Even so, a new technique may not be introduced even when the potential benefits have been perceived, due to an unwillingness to change the present system. At the other extreme entrepreneurial farmers may try out techniques before the necessary development work has been carried out.

The rapid transfer of technology therefore leads to some commercial farms becoming the test-bed for some techniques. This can be an inefficient process if techniques are not retained on farms because they are not fully understood and are incorporated incorrectly. There continues to be a need for a development phase which tests techniques under the complex conditions to be faced on farms, before the incorporation process is encouraged.

Problems caused by inadequate evaluation

Cattle performance not evaluated

In some cases a technique which is potentially beneficial is incorporated into farm systems, even though its effect on physical or financial performance has never been evaluated. This deficiency often occurs with techniques which are intended to measure accurately a parameter, but the parameter may not be important. Two examples are given.

A combination of a walk-through weighing platform sited at the milking parlour exit, and the necessary electronics and computer enables changes in individual cow weight to be closely monitored and analysed (Filby, 1980). Unfortunately, changes in liveweight do not necessarily reflect changes in body fatness (the latter might be a useful parameter on which to base some nutritional management decisions). Changes in body water, in gut fill, and gains in bone and muscle weight prevent precise relationships being determined between fat deposits and liveweight. So, although the development of walk-through weighers for dairy cattle might appear a useful technique, in practice it has little relevance.

A second example is the milk progesterone test for early pregnancy diagnosis in dairy cows. This technique resulted from research into radioimmunoassay measurements of progesterone in blood and milk (Heap et al., 1973; Pope et al., 1976). The milk test carried out 3 weeks after insemination has been shown to be about 85% accurate for cows diagnosed pregnant and almost 100% accurate for cows diagnosed not pregnant. Recent developments of simple on-farm tests have further increased the attractiveness of the test. However, the test is only of use if it leads to an improvement in reproductive efficiency as measured by the herd calving index. Unfortunately, this assessment has never been carried out.

Cattle performance evaluated in isolation

The traditional scientific approach to research is to reduce the question under test to a component comparison. While this provides precision in a relationship, it avoids the interactions which occur when introduced into a larger system, such as a farm.

One example is the relationship between nitrogen (N) fertilizer input and herbage dry matter (DM) production. A large number of experiments on plots harvested by machine have indicated a response of 15–25 kg DM/kg N fertilizer, up to fertilizer levels of about 350 kg N/ha. Advisory recommendations have been made on the basis of these results. Estimates on dairy farm systems where pastures are used for both grazing and cutting indicate however, that a response of only about 8 kg DM/kg N is achieved in practice (Leaver, 1985). The effects of grazing, the return of dung, urine and slurry and the presence of legumes are possible interacting components to explain the poorer farm response.

Another example is the system of concentrate allocation for dairy cows. Research work in the 1950s and 1960s showed clearly that the milk yield response to an increment of feed energy input was greater for high than for low potential cows. Thus, feeding systems continued to evolve which allocated concentrates according to the current yield of the individual cow. The subsequent development in the 1970s of computer-controlled electronic feeders also allowed automatic feeding of individuals to take place. This evolution of complex feeding systems was brought to an abrupt halt when further feeding experiments (with forage fed *ad libitum*) showed no difference in mean milk yields between groups of cows which were fed concentrates either according to yield or flat rate (all cows fed the same each day) (Leaver, 1986). These results were confirmed at different levels of concentrate input, for forages of high and low quality, and for cows in early and mid-lactation. The reasons for the success of the flat-rate system appear to be the compensating effects of forage intake. Under a*d libitum* forage conditions a similar response (kg milk/kg concentrate) is achieved by cows of different yield levels.

Oversimplistic economic evaluation

A new technique or technology should increase economic efficiency and/or simplify the working of the system if it is to appear attractive to a farmer. The financial assessment of a new development is however often oversimplified in the development phase and leads to an inaccurate prediction of the effect on whole farm profit. This is particularly the case when the new system influences the size of the whole farm business.

In the pre-quota dairy industry, a whole range of feeding regimens were practised. Research indicates a mean response of 0.79 kg milk/kg concentrate DM (Broster and Thomas, 1981). At that time the milk price:concentrate price ratio was about 1:1 which simplistically suggested that concentrate feeding was uneconomic. Many farmers were therefore advised to practise low concentrate feeding systems. However, this analysis ignored completely the effect of concentrates on forage intake, and its beneficial effect on stocking rate (Table

Table 15.1 Systems study of high and low concentrate inputs

	High	Low
Grassland area (ha)	32	32
Milk sales/cow (l)	6088	5168
Concentrates/cow (kg)	2244	992
No. of cows	85	69
Relative:		
Margin over feed/cow	100	105
Gross margin/ha	100	86
Profit/ha	100	56

Leaver and Fraser, 1987.

15.1). This study showed that although margin/cow was reduced, farm profit increased with greater amounts of concentrates.

Environmental and social consequences not evaluated

Agriculture in developed countries has expanded output considerably over the past 40 years, and increasing the output of cattle systems (per animal and per hectare) has been profitable for most farmers. This approach is no longer the only way forward, as surplus production has focused attention on reducing costs and increased awareness of environmental and social constraints.

Research into silage making systems for example, has shown that providing a satisfactory additive is used, direct cutting of low DM grass for silage gives as good if not better performance in beef and dairy cattle than wilted silage (Flynn, 1981; Gordon, 1981). Although research workers and extension workers have emphasized the need for adequate storage systems to deal with the silage effluent from direct-cut systems, many farms do not possess adequate facilities. The trend to lower DM silages has therefore been accompanied by an increase of pollution incidents from silage effluent.

There is also greater concern over the loss of nitrogen (N) from intensively grazed pastures, into groundwater, into watercourses and to the atmosphere. Controlled experiments with cut herbage suggest a recovery of applied N in harvested herbage of 62–72% (Morrison, Jackson and Sparrow, 1980). However, farmlet studies, where recycled animal excreta adds substantially to the return of N, indicate a large loss of N from the system (Table 15.2). These losses from intensively managed grassland will in future come under increasing scrutiny.

Table 15.2 Recovery of nitrogen (N) from an intensive grassland unit

	Kg/ha
Fertilizer N	356
Slurry N	107
Dung and urine N	112
Total N applied	575
Total N utilized	245
Nitrogen not recovered	330

Leaver and Fraser, 1989.

Two examples where public concern over animal welfare, and over human health have influenced cattle systems, are veal production and the use of hormone implants. Systems of veal production involving individual crates which restrict movement, and the feeding of fibre and iron deficient diets are no longer allowed under The Welfare of Calves Regulations, 1987. These ensure that calf pens allow calves to turn, and that sufficient digestible fibre and iron are provided in the diet to maintain good health and vigour (Webster, 1988). Similarly the use of hormone implants in beef cattle, which increase growth rate and food conversion efficiency (Steen, 1985), is now banned in EEC countries. The ban was imposed as a result of expressed concern over hormonal residues in meat in spite of scientific evidence to the contrary.

Some of the above problems in incorporating new techniques into cattle production systems are due to research being too narrowly focused. This suggests a strong need for a greater use of multidisciplinary teams of research workers in technique evaluation. This will ensure that all the interacting components in a system are considered. During the development phase, the economic, environmental and social consequences of incorporating new technology must also be examined. A systems approach is therefore an appropriate methodology in developed as well as in developing agricultural systems.

The systems approach

Cattle production systems are ecosystems involving animals, plants and soil interacting with people and with the environment, within a financial framework. This is illustrated in Figure 15.2. Even in this simple form the model shows the complex and dynamic nature of the agroecosystem within which the production system is functioning. The outputs feed back to influence the framework within which the technology system operates. Such a model applies to different types of technology systems in both developing and developed countries, although the nature of the components and the magnitude of the constraints differ enormously.

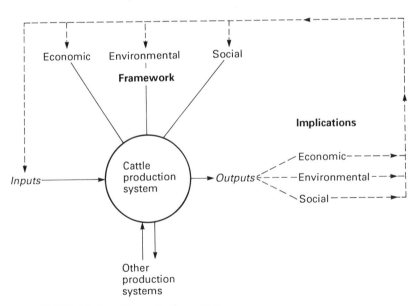

Figure 15.2 Model of a cattle production system

In developed countries, the influence of economic, environmental and social constraints on production systems will mean greater scrutiny of inputs and outputs. An outline of the implications to be considered is given in Table 15.3.

The major economic implication is how a new technique will affect farm profit, through its effects on overhead and variable costs, and on output and product price.

Table 15.3 Some implications to be considered when incorporating new technology into farm systems

Economic

    land requirement
    labour requirement
    capital requirement
    influence on cattle numbers
    influence on variable costs
    influence on output
    influence on price and marketability of product

Environmental

    pollution (from nitrates, ammonia, agrochemicals)
    species diversity (animals and plants)
    landscape features (buildings, trees, hedgerows)
    soil erosion

Social

    human health (animal fat, hormones)
    animal welfare (housing, transport, slaughter)
    rural community (labour, public access)

The financial pressures on farmers have meant greater attention being given to the control of borrowings to reduce interest charges, and to the marketing of the products. The increased sales of low-fat milks, the diversification in dairy products and the reduction of fat levels in beef are examples of increasing market awareness.

The interaction of agriculture and the environment probably represents the greatest constraint to new technologies in agricultural land use. Increasing pollution from animal excreta, silage effluent and agrochemicals, loss of species diversity, loss of landscape features and soil erosion are all of public concern (Green, 1986). Such environmental matters have traditionally been in the hands of farmers, but new technologies and the increasing size of farms have led to a perceived deterioration in the countryside environment, particularly in the large arable areas. Some of these trends are now being reversed partially through the financial incentives of management agreements concerning production systems, and partly through legislative controls.

The social concerns which might influence the incorporation of new techniques into cattle systems can be identified as cattle products and human health, and animal welfare, particularly in relation to housing. More general issues include the influence of a changing agriculture on rural communities and public access to agricultural land.

Conclusions

In developed countries, financial incentives plus the support of strong research and development programmes and extension agencies, have led to substantial increases in output and efficiency of cattle production systems. The resulting production surpluses have brought this era of greater total output to an end. Nevertheless, improvements in animal production efficiency including in some cases greater output per animal will remain an objective of new technologies. It is essential that farms remain financially viable otherwise the rural environment will deteriorate. Thus, the increased efficiency of agricultural systems remains a priority, and alternative more holistic strategies are required.

The exciting new techniques being generated by research in breeding, feeding, housing, mechanization, and health control systems require continuing development, and where appropriate, incorporation into cattle production systems. Also it is essential that there is continuing support for agricultural research if the industry is to remain thriving and competitive.

The environmental and social implications for new techniques which have been highlighted, represent an opportunity and challenge to those in research, development and advisory work, and this is a challenge already being met by many progressive farmers. The appreciation that new techniques introduced into farm systems, influence other components within the system and have environmental, social as well as economic implications should lead to more interdisciplinary research programmes and to a greater amount of modelling and testing of farm systems.

References

Broster, W. H. and Thomas, C. (1981) The influence of level and pattern of concentrate input on milk output. In *Recent Advances in Animal Nutrition – 1981*, (ed. W. Haresign). Butterworths, London, pp. 49–70

Filby, D. E. (1980) Cattle weighing. In *The Mechanisation and Automation of Cattle Production*. Occasional Symposium No. 2. British Society of Animal Production, Reading, pp. 25–36

Flynn, A. V. (1981) Factors affecting the feeding value of silage. In R*ecent Advances in Animal Nutrition – 1981*, (ed. W. Haresign). Butterworths, London, pp. 81–90

Gordon, F. J. (1981) The effect of wilting of herbage on silage composition and its feeding value for milk production. *Animal Production,* **32**, 171–178

Green, B. H. (1986) Agriculture and the environment: a review of major issues in the UK. *Land Use Policy,* **3**, 193–204

Heap, R. B., Gwyn, M., Laing, J. A. and Walters, D. E. (1973) Pregnancy diagnosis in cows; changes in milk progesterone concentration during the oestrous cycle and pregnancy measured by rapid radioimmunoassay. *Journal of Agricultural Science, Cambridge,* **81**, 151–157

Leaver, J. D. (1985) Milk production from grazed temperate grassland. *Journal of Dairy Research,* **52**, 313–344

Leaver, J. D. (1986) Systems of concentrate distribution. In *Principles and Practice of Feeding Dairy Cows*, (ed. W. H. Broster, R. H. Phipps and C. L. Johnson), Technical Bulletin 8, NIRD, Reading

Leaver, J. D. and Fraser, D. (1987) A systems study of high and low concentrate inputs for dairy cows: physical and financial performance over four years. *Research and Development in Agriculture,* **4**, 171–178

Leaver, J. D. and Fraser, D. (1989) A systems study of high and low concentrate inputs for dairy cows: grassland production and utilisation over four years. *Research and Development in Agriculture,* (in press)

Morrison, J., Jackson, M. V. and Sparrow, P. E. (1980) The response of perennial ryegrass to fertiliser nitrogen in relation to climate and soil. *Report of the Joint GRI/ADAS Grassland Manuring Trial GM20*, Technical Report 27, The Grassland Research Institute, Hurley

Pope, G. S., Majzlik, I., Ball, P. J. H. and Leaver, J. D. (1976) Use of progesterone concentrations in plasma and milk in the diagnosis of pregnancy in domestic cattle. *British Veterinary Journal,* **132**, 497–506

Steen, R. W. J. (1985) A comparison of bulls and steers implanted with various growth promoters in a 15-month semi-intensive system of beef production. *Animal Production,* **41**, 301–308

Webster, A. J. F. (1988) Beef cattle and veal calves. In *Management and Welfare of Farm Animals*. The UFAW Handbook, Bailliere Tindall, London, pp. 47–80

Chapter 16

The incorporation of new techniques into cattle production systems in developing countries

M. B. Aboul-Ela and A. S. Abdel-Aziz

Developing countries comprise about 58% of the world land area, on which about three-quarters of the total world population live. Cattle production in developing countries is surpassed by that in developed countries. Although the total cattle population (both cattle and buffaloes) in developing countries represents about 70% of the world population, they produce only 29 and 23% of the world's cattle-meat and milk respectively (Table 16.1). These figures reflect the higher

Table 16.1 Land use, human population, cattle and buffalo population and productivity of milk and meat in developed and developing countries

	Developed countries	Developing countries	Total
Land use			
Land area (ha)	5 488 592	7 592 422	13 081 014
Arable and permanent crops (ha)	676 544	800 217	1 476 761
Permanent pasture (ha)	1 257 204	1 894 133	3 151 337
Population (in 1000s)			
Total population	1 210 437	3 626 523	4 836 960
Active population	580 255	1 583 490	2 163 745
% of active population in agriculture	10.6	63.0	49.0
Cattle population (1000s)	423 145	845 789	1 268 934
Buffalo population (1000s)	727	128 556	129 283
Total (cattle + buffalo)	423 872	974 345	1 398 217
Milk production (fresh milk, tonne $\times 10^3$)			
Cattle	379 699	78 324	458 023
Buffalo	81	32 422	32 503
Total (cattle + buffalo)	379 780	110 746	490 526
Meat production (tonne $\times 10^3$)			
Cattle	33 283	12 786	46 069
Buffalo	12	995	1 007
Total (cattle + buffalo)	33 295	13 781	47 076

FAO (1985a).

productivity levels for cattle in developed countries which have been achieved through comprehensive development programmes and implementation of new technologies over the past 50 years. These values may, however, be taken to indicate that there is wide room for future development of cattle production in developing countries.

Developing countries extend over a wide variety of ecological zones, and social and political structures which make it extremely difficult to discuss in general terms aspects of technology incorporation. Cattle production systems in developing countries range from the pastoral and transhumant, as in parts of Africa and Latin America, to small-holdings with mixed farming (crops, small ruminants and large ruminants), usually in settlements and irrigated areas. They also include the relatively less frequent and more recent specialized systems of cattle production as in large dairy units in Egypt and the Gulf states. With the exception of the latter systems, cattle breeds in developing countries cannot be easily characterized as dairy or beef breeds. They are mostly used as dual purpose animals, and in some cases they are used also as draft animals. Because of this variation in ecological conditions and production systems, the common factors that affect the incorporation of new technologies in these countries are considered.

Constraints to incorporating new technologies into developing countries

New technologies can be classified into 'hard technology' including machinery, improved genotypes etc. and 'soft technology' which includes methodologies and the 'know-how'. There are many constraints to the success of new technology in developing countries which can be categorized into infrastructural, environmental, financial, social and cultural.

Infrastructural constraints

The development of modern cattle production at a national, or regional level is dependent on the adequacy of the main infrastructural services, e.g. road system, provision of electricity and power supply and institutional services. For example, the development of cattle production in Sudan and other parts of Africa is limited by the inadequacy of road and transportation systems for the transfer of both the input of materials and the removal of products. Under such conditions also, the conventional artificial insemination (AI) networks currently used in western Europe would be almost impossible to apply. Needless to say, without a reliable power supply, hard technology, such as milking machines, computers and AI equipment etc. cannot be incorporated into cattle production systems.

The lack of developed organized marketing systems in developing countries, particularly for milk products, constitutes another major constraint.

Climatic constraints

Most of the developing countries are in the tropical and subtropical regions where climatic factors hinder animals' productivity (Shafie, 1989). Adverse effects of climate on productivity have been clearly demonstrated for native breeds in developing countries. They are even more pronounced for exotic breeds of imported genotypes which are imported into developing countries to be used on small-holdings or large enterprises. A clear example of this is the very low performance of Holstein cattle raised in large modern farms in Saudi Arabia near Riyadh where fertility levels have been reported to be exceptionally low (Ryan *et al.*, 1988) coupled with reduced milk production. Under these conditions, heat stress affects the success of the use of many technologies and makes them inappropriate to these conditions. Specific new technology (spray cooling) may have to be designed for these conditions to ameliorate heat stress, but these need to be cost-effective for the whole system of technology transfer to survive.

Feed resources

Seasonal variation of feed resources in most developing countries is a key determinant for the production system that has to be followed. The feed resources available are also often of low quality which hinders cattle productivity even when new technologies are introduced. An example of this is the recent changes in the composition of concentrate feed mixture (manufactured centrally by local authorities) in Egypt. Replacing part of the cottonseed cake with roughage material has reduced the total digestible nitrogen (TDN) value of the mixture, which cannot now meet the nutrient requirements of imported high yielding dairy cattle, even at high intake levels. Consequently, the new technology introduced in the form of improved genetic resources has become unsuccessful.

Under extensive natural range conditions, over-grazing and misuse of pasture areas, particularly near water points and along cattle routes, worsens the situation of feed shortage. In addition, range conservation and development is seldom practised.

Farm and herd size

With the exception of some Latin American and a few African countries where large extensive beef cattle operations exist, most of the developing countries are characterized by small-holdings. According to a recent survey in India (El-Ghonemy, 1984), 14% of the livestock producers are landless and 38% are small-holders. Available statistics indicate much smaller holdings areas in Africa and Asia (median area of 1–5 ha in Africa and often less than 10 ha in Asia) than in north and central America (median size ranging from 40 to 500 ha) and the EEC countries (>15 ha on average) (FAO, 1984). Similarly, herd size in most African and Asian developing countries is much smaller than in EEC and North American countries. Recent surveys in Egypt by Nigm *et al.* (1986), Abdel-Aziz (1987) and El-Keraby (personal communication) indicate that herd size of both cattle and buffaloes in the Nile delta region has a mode of one to four. In the EEC countries, about 50% of the herds are of over 10 cows, with average herd size of 16.5 heads, the highest figures (58 heads on average) recorded for the UK (EEC, 1985).

The main drawback of small herds is the lower ability of their owners to provide the capital input necessary for acquiring new technologies which are designed mainly for larger operations. This is particularly true for 'hardware' (e.g. milking machines, computers for recording and monitoring systems etc.) and other inputs necessary for the process of intensification (e.g. feed additives and concentrates), and facilities required for better marketing processes (packing, transport etc.). Furthermore, new technologies may require the employment of skilled technical personnel whose presence will be cost-effective only in large operations. Despite these disadvantages of small herd size in developing countries, productivity levels under these conditions do not seem to be much less than those obtained in large state farms. In fact, recent on-farm studies in Egypt indicate even higher levels of milk production (Nigm *et al.*, 1986; Abdel-Aziz, 1987) and better reproductive performance (Aboul-Ela, 1989) of buffaloes in small-holdings than those reported for large farms. This has been attributed to the better management and care given to individual animals in small herds.

Since it is well recognized now that small herds will remain an essential feature of the rural sector in developing countries, long-term development programmes, which involve the introduction of new technology, have to be adopted to suit this situation. The results of such programmes so far, have been mixed. Technology transfer under these circumstances should be sought within a complete package that would take care of many of the input and output elements of the production system in a particular environment.

Projects that involve the provision of new technology, through cooperatives or central governmental authorities, may be a better solution in the small-holding system.

Financial constraints

Levels of income in developing countries are generally lower than in developed countries. Accordingly, the price of unit output in developing countries does not, in many cases, justify an economic introduction of new technology, the basic components of which are often produced in developed countries and sold at a relatively high price. Unless subsidized by local governments or by the producing developed countries, cattle producers would not be able to afford to use such new technology. An example of this is the use of AI. When based on milk prices, for fair comparison, an insemination in the UK using the 'bull-of-the-day', costs the farmer the equivalent of about 10 litres of milk. Total unsubsidized cost of an insemination in Egypt is equivalent to about 20 litres of milk. Unless the costs of such services are reduced through subsidy, or through local production of its major components (e.g. straws, semen, containers etc., as in India) farmers would find it difficult to afford such services. Another example is the use of hormonal treatment for oestrous synchronization, the cost of which in developing countries would be at least three or four times that in developed countries. The new technology has to be cost effective for its successful transfer into developing countries.

It is worth noting that most studies indicate that economic incentives are necessary to promote changes acquired through new technology. Insurance for innovators, systems of fair pricing of products and services are essential, while subsidies for inputs are generally justified in the early stages of development.

New technologies are often more readily adopted by large producers than small-holders. The former usually have easier access to information and inputs, face lower transaction costs per productive unit, can better afford the risk and above all have the financial capacity to acquire the 'hardware' of the new technology. In the case of small-holders in developing countries, this may be overcome through the provision of new technology by the government or local authorities using cooperatives. Having a credit system has been proved in many cases to be beneficial, particularly as soft loans to establish or generate cattle enterprises.

The success in providing central governmental services or establishing credit systems is limited, however, by the financial problems of the governments themselves, particularly as the foreign exchange component of the gross investment is often more than 20%, as estimated by FAO, for all levels of technology (Gifford, 1984). Many of the developing countries have continued to face serious debt-servicing difficulties. According to the World Bank estimates, the external indebtedness of developing countries rose from US$610 billion in 1980 to US$970 in 1985 which, combined with high interest rates, further increased the debt servicing burden (FAO, 1985b). This situation has forced some of the developing countries to cut back their expenditure on new projects and also forced international lending organizations to cut down their lending commitments. This, undoubtedly, will have a negative effect on the transfer of new technology, particularly those planned through governmental organizations.

A further point which is worth considering is that many technology transfers in the past have failed because of lack of foreign exchange to buy spare parts which has left equipment idle. This has to be seriously considered in future planning for developing countries.

Social constraints

It is well recognized that there is a need to reckon with sociocultural and behavioural factors when attempting to incorporate new technology in developing countries. Magical and religious beliefs, and certain aspects of the social system such as cast, gerontocracy, low status of women, mutual distrust can be serious obstacles to rational economic decisions, and hence to incorporation of new technologies. For example, recording systems which are the corner-stone in any genetic improvement programme have not taken off in some developing countries because of the opposition of small farmers, due to social beliefs or sheer envy.

In developing countries a major sector of the population is engaged in agriculture (see Table 16.1). The introduction of new technology usually aims at intensifying the use of resources and often calls for the replacement of traditional labour-demanding methods by advanced technology. The introduction of new technology has to be effective enough to increase productivity so that additional employment is generated.

In some rural areas, the size of the herd plays a major role in establishing the social position of individuals or families regardless of their productivity or economic efficiency. Costs of many social events (e.g. dowry) are paid in cattle rather than money.

Level of education and training

General education contributes to improved animal production by facilitating the transfer of technological knowledge, improving receptiveness to new ideas and by changing the values and aspiration of farmers. Although data on adult literacy in developing countries are limited, the level of illiteracy in parts of Africa and the Near East (about 65%) particularly in rural areas is much higher than in developed countries (El-Ghonemy, 1984). It has been claimed that low levels of literacy may not prevent farmers from adopting new technology (Arnon, 1987), yet with the increasing levels of sophistication in agriculture, the lack of education can become a constraint to further progress. In genetic improvement programmes, which are based mainly on recording systems, illiteracy and low levels of general education are considered a real constraint, not only because the farmers cannot perform the actual recording but they also do not appreciate the value of keeping records for their herds. This is true also for other technologies, as in the use of hormones for oestrous synchronization and the implementation of diseases control programmes. In such cases, extension has a much more important role to play, compared to that in developed countries, particularly for small farmers. On the other hand, large scale operations run either as state-owned or private enterprises usually employ workers who are able to acquire and absorb the new technology.

Lack of trained skilled personnel for the maintenance and servicing of new machinery often causes serious problems to the success of technology transfer. Thus, provision of training should be considered as an important component in any development programmes for developing countries.

Genetic resources

As indicated earlier, productivity level of cattle population in developing countries is much less than that in developing countries, partly due to low genetic potential. There is a lack of comprehensive planning for genetic improvement either by selection or by replacement to influence a major proportion of the national cattle population. Quite often, the introduction of temperate breeds to subtropical and tropical countries has had limited success due to unfavourable environmental conditions. These have to be considered when evaluating the economic efficiency of the new technology introduced.

Appropriate technology or new technology?

Due to wide variation in the production system and its different components between developing and developed countries, it is wrong to assume that new technology introduced into developed countries is appropriate for developing countries. Appropriate technology, in the context of cattle production refers to the level of technology which is best suited for introduction and use in a specific development situation. Appropriateness of a technological input is determined by the technical, economic, social and political characteristics of each development situation. Therefore, it is not possible to generalize on the appropriateness of a particular level of technology, nor is it possible only to examine one particular item of technology and decide if it is appropriate.

The main requirements of appropriate technology are:

(1) technical feasibility; it must be capable of increasing productivity given the necessary technical elements
(2) economic feasibility; it must be cost-effective with a risk level which a farmer can afford
(3) compatibility with the farming system; the technology must have requirements that fit in with the farming system
(4) social acceptability; it must be compatible with the norms and beliefs of the community
(5) safety to the farmer or the user to cope with their low level of education
(6) infrastructure compatibility; it must have requirements which can be accommodated easily by the existing level of infrastructure.

In many instances an 'appropriate' technology may have to be 'tailored' for a particular situation. Examples of these are: developing small tractors to fit in small-holdings, mobile milking machines or small dairy processing units to be shared by more than one small farmer, use of solar-powered equipment whenever electric power is not generated. It is worth emphasizing that a new technology may be sought within a package of resources to ensure its success.

References

Abdel-Aziz, A. S. (1987) Performance of dairy cattle and buffaloes in small holdings. In *Proceedings of the Egyptian-Austrian Symposium on Dairy Cattle Husbandry*. 26 February, 1987, Cairo. Paper no. 5 International Egyptian Centre for Agriculture, Cairo

Aboul-Ela, M. B. (1989) Reproductive patterns and management in the buffalo. In *Proceedings of International Symposium on Constraints and Possibilities of Ruminant Production in the Dry Subtropics*, Cairo 2–4 November 1988. Pudoc, Wageningen, The Netherlands, pp. 174–79

Arnon, I. (1987) *Modernization of Agriculture in Developing Countries: Resources, Potentials and Problems*. 2nd edn. John Wiley & Sons, Chichester, UK

EEC (1985) *Dairy Facts & Figures 1985*. Milk Marketing Board, Surrey, UK

El-Ghonemy, M. R. (1984) Development strategies for the rural poor. *FAO Economic and Social Development Paper No. 44*. Food and Agriculture Organization of the United Nations, Rome

FAO (1984) Agricultural holdings in the 1970, World census of agriculture: a statistical analysis. *FAO Economic and Social Development Paper No. 47*. Food and Agriculture Organization of the United Nations, Rome

FAO (1985a) FAO Production yearbook Vol. 39. *FAO Statistics Series No. 70*. Food and Agriculture Organization of the United Nations, Rome

FAO (1985b) Commodity review and outlook. *FAO Economic and Social Development Series No. 40*. Food and Agriculture Organization of the United Nations, Rome

Gifford, R. C. (1984) Agricultural mechanization in development: guidelines for strategy formulation. *FAO Agricultural Services Bulletin No. 45*. Food and Agriculture Organization of the United Nations, Rome

Nigm, A. A., Soliman, I., Hamed, M. K. and Abdel-Aziz, A. S. (1986) Milk production and reproductive performance of Egyptian cows and buffaloes in small livestock holdings. In *Proceedings 7th Conference of Animal Production*, Cairo, 16–18 September, Egyptian Society of Animal Production, Cairo, pp. 290–304

Ryan, D. P., Kopel, E., Boland, M. P. and Maceoin, F. (1988) Effects of receptal administered at time of AI or day 12 of the oestrous cycle on pregnancy rates in Holstein cows in Saudi Arabia. In *Proceedings of 11th International Congress on Animal Reproduction and Artificial Insemination*, 26–30 June 1988, Dublin, Vol. 4, paper 457

Shafie, M. M. (1989) Environmental constraints on animal productivity. In *Proceedings of International Symposium on Constraints and Possibilities of Ruminant Production in the Dry Subtropics*, Cairo, 2–4 November, 1988. Pudoc, Wageningen, The Netherlands, pp. 10–16

Chapter 17

Short descriptions of specific new techniques

Nutrition

Complementary feeding of grass silage and compound feed to dairy cows

T. D. A. Brigstocke[1], R. D. Gill[1] and C. P. Freeman[2]

The rates at which silages ferment in the rumen are a function of the content of readily available carbohydrates and the degradability of the polysaccharide fractions of the carbohydrate complex. A fermentation index (FI) has been described, defined as the content of rapidly-degradable hemicellulose and soluble carbohydrates, together with lactic acid expressed as a proportion of silage dry matter (DM). When the FI of a silage is high, conditions of low (and unstable) rumen pH, reduced ruminal outputs of volatile fatty acids and low ruminal acetate:propionate ratios are likely to result, which could lead to reduced silage intakes and lower outputs of milk and milk constituents. The effect on dairy cow performance of modifying the composition of concentrates to alleviate these silage-induced rumen conditions has been examined.

Materials and methods

Experiment 1

Twenty-four Friesian dairy cows in early lactation, paired by calving date and previous yield, were individually fed silage (269 gDM/kg, 10.4 MJ ME/kg DM, 181 gCP/kg DM, pH 4.1, FI 24) and either modified (complementary) or unmodified (control) concentrate according to milk yield for 8 weeks.

The complementary concentrate was formulated to have a reduced content of readily-available carbohydrate, increased levels of slow-fermentable starch, and increased buffering capacity relative to the control concentrate. Metabolizable energy (ME) and crude protein (CP) contents of the complementary and control concentrates were 14.0 and 13.9 MJ/kg DM, and 205 and 198 g/kg DM respectively. Silage intakes and milk yields were recorded daily and milk compositional analysis carried out at 2-weekly intervals.

[1] BOCM Silcock Ltd, Basing View, Basingstoke, Hants, UK.
[2] Unilever Research, Colworth House, Sharnbrook, Bedford, UK

Experiment 2

The effects on silage intake of feeding modified or unmodified concentrates with Italian ryegrass silage (317 g DM/kg, 10.8 MJ ME/kg DM, 140 g CP/kg DM, pH 4.0, FI 3.1) were examined for 3 weeks using 24 individually-fed Friesian dairy cows. The complementary, modified as described above, and control concentrates were fed at 8 kg/day and had ME levels of 12.2 and 12.0 MJ/kg DM and CP contents of 194 and 200 g/kg DM respectively. Cows were fed either silage alone or together with complementary or control concentrates.

Results and discussion

Experiment 1

Concentrate intake tended to be greater for complementary (9.4 kg/day) than control (8.8 kg/day) concentrates, and silage DM intake was significantly greater for cows fed complementary concentrate (Table 1). These cows also tended to have

Table 1 Effect of concentrate type on silage DM intake and milk production. Experiment 1

Concentrate	Silage DM intake (kg/day)	Milk yield (kg/day)	Fat yield (kg/day)	Protein yield (kg/day)	Fat content (g/kg)	Protein content (g/kg)
Control	9.8	27.0	1.05	0.80	38.1	29.0
Complementary	10.6	28.9	1.08	0.87	36.7	29.3
Significance	+	NS	NS	NS	NS	NS

NS: Not significant. $+P<0.05$

increased milk yield and milk protein yield, and decreased milk fat content, although these differences were not significant.

Experiment 2

Silage DM intake of cows fed no concentrate, control concentrate and complementary concentrate were 13.1, 10.6 and 12.0 kg/day respectively (SE ± 0.83). There was therefore a trend ($P>0.05$) for the substitution effect to be reduced with the complementary concentrate.

Conclusion

The alleviation of the rumen effects of high FI silages by feeding appropriate complementary concentrate tended to produce beneficial effects on both silage intake and milk production. An important part of this approach to complementary feeding is the identification of silages which fall into a rapidly-fermenting category.

Supplementation of dairy cows with yeast culture

J. A. Bax[1]

The effects of supplementing grass silage and concentrate based diets with a viable yeast supplement were studied with 12 British Friesian cows. They were assigned to four treatments:

(1) S: 5 kg concentrate and silage *ad libitum*
(2) SY: as for S but with yeast culture added
(3) C: a complete mix offered *ad libitum* containing 0.25 concentrate and 0.75 silage on a dry matter (DM) basis
(4) CY: as for C but with yeast culture added.

When a highly concentrated yeast culture (Alltech Inc, Lexington) has been used as a feed supplement in dairy cow rations increases in both yield of milk and yield of milk constituents have been reported (Boland, 1986; Hoyos, Garcia and Medina, 1987). The yeast culture is based upon a strain of *Saccharomyces cerevisiae* and a growth medium which is preserved in a viable form.

Method

Twelve individually fed cows (second lactation or more) were allocated to a 9-week changeover experiment. The four treatments are described above. The cows on treatments S and SY were offered their concentrate in two separate out of parlour feeds daily. The concentrate contained 783 kg barley, 188 kg soya and 29 kg minerals/tonne DM. The yeast culture was added to the minerals as a premix at the rate of 2 kg/tonne concentrate mix, freshweight. The chemical composition of the feed is given in Table 1.

Table 1 Chemical composition of the feed (g/kg DM) unless otherwise stated

	Silage	*Concentrate + Yea–Sacc*	*Concentrate – Yea–Sacc*
Oven DM (g/kg)	182	835	838
Organic matter	913	949	948
Crude protein	161	186	190
In vitro digestibility	68.2	80.1	79.7
ME (MJ/kg)	10.8	12.8	12.7
Ammonia nitrogen (g/kg total N)	85	–	–
pH	3.8	–	–
NDF	536	138	149
ADF	333	63	67

ME: metabolizable energy; NDF: neutral detergent fibre; ADF: acid detergent fibre.

Results

The addition of the yeast culture to the diets did not significantly affect milk yield or silage DM intake, although both tended to be slightly increased, particularly

[1] West of Scotland College, UK

Table 2 Feed intake and milk production

	Treatment				
	S	SY	C	CY	SED
Silage (kg DM/day)	9.45	9.92	10.29	10.40	0.24
Concentrate intake (kg DM/day)	4.93	4.93	3.55	3.60	0.052
Milk yield (kg/day)	18.2	18.8	17.1	17.4	0.43
Milk composition (g/kg)					
fat	42.4	42.9	43.6	42.8	0.99
protein	33.4	33.3	32.2	32.5	0.30
lactose	45.8	45.8	45.4	46.4	0.40
Yield (g/day)					
fat	770	807	746	744	21.8
protein	607	629	551	557	15.9
lactose	836	868	780	808	24.8

when the silage and concentrates were fed separately (Table 2). There were also no significant effects of adding yeast culture to the diet on fat, protein or lactose content in the milk for cows fed either a complete diet or silage and concentrates separately.

The cows offered the complete diet received less concentrate and ate more silage DM than the cows fed silage and concentrate separately. They produced less milk of similar fat and lactose concentration but lower protein concentration.

Conclusion

It is concluded that small or no benefits are likely to arise from the inclusion of yeast culture in a grass silage and concentrate diet for dairy cows.

References

Boland, R. (1986) Effect of Yea-Sacc on dairy herd performance. *Annual Report.* Dordt College
Hoyos, G., Garcia, L. and Medina, F. (1987) Effects of feeding viable microbial feed additives on performance of lactating cows in a large dairy herd. American Dairy Science Association Meeting, Columbia, 1987, abstract

The effect of a high-strength yeast culture in the diet of early-weaned calves

J. Hughes[1]

The use of biological food additives in animal production systems is becoming common, although published data are limited. Yea-sacc, a high-strength yeast culture, is one such product. The yeast used is a selected strain of *Saccharomyces cerevisiae*, and the dried culture contains this, and the medium on which it is grown.

A trial was conducted to examine the effect of including this yeast culture in the diet, on the performance of early-weaned calves. Two groups of 16 calves (eight dairy, eight beef cross) aged 1 week with a mean liveweight of 52 kg were allocated to one of two dietary treatments: C, control and Y, yeast culture (included in the concentrate portion of the diet, at 2 g/kg in the starter, and 1 g/kg in the rearer). Calves were bucket-fed once daily for 35 days before being abruptly weaned. The starter pellets (207 g crude protein (CP)/kg dry matter (DM)) were fed *ad libitum* for 42 days, followed by rearing nuts (172 g CP/kg DM) which were fed *ad libitum* for a further 42 days. The diets were proprietary compounds containing several cereal and protein sources. Straw and water were freely available. The trial lasted 84 days. Yeast culture inclusion significantly increased daily liveweight gain ($P<0.05$): 0.737 and 0.831 (s.e.d. 0.0461) kg/day for C and Y respectively. There was no difference between breed types. Y calves consumed significantly more food to weaning ($P<0.05$), although there was no difference over the 84-day period. Calves were sold at the end of the trial, with the heavier Y calves worth more. After deducting the extra food cost for the Y group, there was still a net extra income over the C group.

The results indicate that inclusion of yeast culture in a calf diet improves both physical and financial performance.

The influence of physical form of maize grain and silage inclusion on the utilization of finishing diets by beef cattle

H. J. Van der Merwe[2], G. Jordaan[2] and W. A. Kottler[2]

Contradictory results were obtained when the influence of the physical form of maize, in finishing diets for beef cattle, was investigated. It would appear that the digestion of whole grain is influenced by factors such as the quantity and physical form of roughage included in the diet. Maize silage is often used as a roughage source in finishing diets for beef cattle. In the case of large feedlots, the purchasing or production of enough silage usually becomes a problem. The cost of roughage and milling of grain could probably be eliminated by including whole grain in an all concentrate finishing diet for beef cattle.

The effect of replacing maize meal (4 mm sieve) with rolled (6 grooves/30 mm) and whole maize grain respectively, in a finishing diet for beef cattle containing 20% maize silage on a dry matter (DM) basis, was investigated. Furthermore, the inclusion of whole maize grain in a diet without silage was also investigated.

[1] Carrs Farm Food Ltd, Carlisle, UK

[2] Department of Animal Science, University of the Orange Free State, Bloemfontein, South Africa

Method

Four groups each consisting of five 2-year-old Angus steers were used. The diets containing approximately 13% crude protein on a dry matter basis were fed *ad libitum*. Silage containing diets consisted (g/kg freshweight) of 470 maize grain; 360 silage; 40 bagasse; 16 poultry litter; 99 liquid feed; 8 limestone and 7 salt. In the case of the all-concentrate diet with whole maize, the composition (g/kg freshweight) was 840 maize grain; 145 liquid feed; 8 limestone and 7 salt. The ingredients (g/kg freshweight unless otherwise stated) of the commercial liquid feed were 330 moisture; 310 crude protein (260 from urea); 6 phosphorus; 26.5 potassium; 31508 IU/kg of vitamin A; 236.1 ppm of monensin-sodium and 78.8 ppm of tylosin. Approximately 10% liquid feed, on a dry matter basis, was included in all the experimental diets.

All the steers were fed individually and slaughtered after a 90-day feeding period. A digestibility study using the same steers was carried out at the end of the finishing period.

Results

Maize meal and rolled maize, compared to whole maize, had a significantly ($P<0.01$) reduced apparent digestibility of crude fibre but a significantly ($P<0.05$) increased apparent digestibility of nitrogen-free extract and metabolizable energy (ME) content. An increase in the apparent digestibility of dry matter, crude protein and nitrogen-free extract as well as ME content was observed when whole maize was included in a diet without silage ($P<0.01$).

With increasing coarseness of the grain, the DM intake of the steers tended to decrease ($P>0.05$). The substitution of maize meal and rolled maize with whole maize in the silage containing diet, resulted in a significantly ($P<0.01$) reduced ME intake. A reduced ($P<0.01$) DM intake was observed when whole maize was included in a diet without silage. No significant difference ($P>0.05$) in the ME intake of steers occurred between the two whole maize diets. A non-significant ($P>0.05$) reduction in live mass gain and efficiency of feed conversion was observed with an increase in the coarseness of the grain. The feeding of whole maize without silage resulted in a non-significant ($P>0.05$) increase in live mass gain and a significantly ($P<0.01$) better feed conversion. No significant ($P>0.05$) differences in carcass mass, dressing percentage and grading occurred between the various treatments.

From the results it was concluded that if the losses ($\pm5\%$) which occurred during the milling of maize are taken into account, it would appear that rolled maize, compared to maize meal and whole maize, is the most efficient physical form in which maize should be fed to steers receiving finishing diets containing 20% silage in the total DM. The feeding of whole maize appears to be justified when an all-concentrate finishing diet is fed to steers, and the most efficient feed conversion was observed with this treatment.

Assessment techniques for feed quality and intake

Development of a method for the *in vitro* estimation of digestibility of forage in ruminants

H. M. Omed[1], R. F. E. Axford[1], A. G. C. Chamberlain[1] and D. I. Givens[2]

Method

The object was to test the accuracy and precision of a simplification of the method of estimation *in vitro* of forage digestibility using faeces liquor (El Shaer *et al.,* 1987). Faeces liquor was prepared by homogenizing 60 g fresh sheep faeces in artificial saliva (pH 6.8) passed through carbon dioxide gas, adding 0.9 g urea, and adjusting the volume to 1 litre. Twenty millilitres of this liquor were added to duplicate 180 mg samples of oven dried herbage of known digestibility *in vivo*, in McCartney bottles. The bottles were closed with screw caps and incubated in the dark at 39°C with frequent shaking for 36 h. The centrifuged residues were incubated with 20 ml acid-pepsin for a further 36 h. The residues were collected by centrifugation, dried and their weight losses recorded as digestibility *in vitro*.

Results

Table 1 shows the precision of the method as given by the standard deviations (SD) of a replicated determination, and its accuracy as given by regression (R)SD of the regression relating estimates *in vitro* to determined digestibility *in vivo*.

Table 1 Relation between *in vitro* and *in vivo* digestibilities (%) of perennial ryegrass and lucerne

Herbage	n	SD	Slope	Intercept	RSD
Perennial ryegrass	(9)	1.42	0.900	8.99	2.26
Lucerne	(8)	1.13	0.958	2.24	0.74
Perennial ryegrass and lucerne	(17)	1.29	0.987	0.51	2.02

SD: Standard deviation; RSD: regression standard deviation.

These results show that the modified method gave satisfactory estimates of digestibility. Incubation in faeces liquor in closed vessels was clean, convenient and easy to accomplish and avoided the need for fistulated animals and expensive chemicals.

Reference

El Shaer, H. M., Omed, H. M., Chamberlain, A. G. and Axford, R. F. E. (1987) *Journal of Agricultural Science, Cambridge*, **109**, 257–259

[1] School of Agricultural and Forest Sciences, University of Wales, Bangor, UK
[2] ADAS Feed Evaluation Unit, Drayton, UK

Prediction of the organic matter digestibility of grass silage

M. S. Kridis[1], G. D. Barber[1], N. W. Offer[1], D. I. Givens[2] and I. Murray[3]

Organic matter digestibility (OMD) is an important aspect of feed quality used to calculate metabolizable energy (ME) for use in ration formulation. Measurement of OMD and ME *in vivo* is both expensive and slow and needs special facilities; however, these values may be predicted provided a suitable *in vivo* calibration population is available. In the UK, several laboratories have conducted evaluations *in vivo* of grass silages since 1970. A total of 170 silages from four populations were used in this work to compare prediction methods.

Method

Table 1 shows the methods used in this study. To ensure satisfactory analytical performance, each method was performed at a single laboratory which uses it routinely, 122 silages were used for the comparative study and the remaining 48

Table 1 Regression statistics for the prediction of silage OMD

	MADF	*LIGA*	*PCOMD*	*NB48*	*IVOMD*	*NIR*
Calibration set n = 122						
R^2	0.34	0.52	0.55	0.68	0.74	0.85
RSD	5.1	4.4	4.2	3.6	3.2	2.5
Validation set n = 48						
R^2	0.20	0.14	0.4	na	0.64	0.76
SEP	5.1	5.3	4.7	na	3.6	2.6
Bias	−0.59	1.18	−2.33	na	−1.85	−0.79

na: data not available.
MADF: modified acid detergent fibre (Clancy and Wilson, 1966).
LIGA: acetyl bromide lignin (Morrison, 1972).
PCOMD: pepsin-cellulase OMD (Jones and Hayward, 1975).
IVOMD: OMD *in vitro* (Alexander and McGowan, 1966).
NB48:   48 h nylon bag rumen incubation (Kridis, Offer and Barber, 1989, *see* p. 229).
NIR: near infra-red reflectance spectroscopy (Norris *et al.*, 1976).

were used to validate the calibration equations. For each method, analysis of variance was used to detect significant differences between population regression lines. Where no significant difference between populations was detected ($P=0.05$) a single line was used to describe all data.

Results and discussion

The overall results are summarized in Table 1. MADF and LIGA gave regression lines which differed significantly in intercept ($P<0.001$) between populations. For

[1] West of Scotland College, Auchincruive, Ayr, UK
[2] Feed Evaluation Unit, Stratford-upon-Avon, Warwickshire, UK
[3] North of Scotland College of Agriculture, Aberdeen, UK

PCOMD, the regression statistics for the individual populations were more precise than those for the other methods. However, the regression lines between populations differed significantly ($P<0.001$) in both intercept and slope and the overall population statistics were poor. The methods which use rumen microbes (NB48 and IVOMD) both described all populations by single regression lines. NIR also gave a single regression line provided that more than five terms (wavelength segments) were used in the multiple regression equation. The best prediction of OMD *in vivo* of a blind test set (n=48) was obtained using an NIR eight-term equation.

Conclusion

For routine advisory application, feed evaluation methods must be rapid, cheap, accurate and easy to operate. Of the methods tested, only the NIR technique meets these requirements. If an NIR instrument is not available, the MADF, LIGA and PCOMD techniques are not satisfactory alternatives, as both precision and accuracy are shown to be poor. The rumen liquor techniques give acceptable precision and accuracy but access to cannulated animals is essential.

References

Alexander, R. H. and McGowan, M. (1966) *Journal of the British Grassland Society,* **21**, 140–147

Clancy, M. J. and Wilson, R. K. (1966) *Proceedings of the Xth International Grassland Congress,* Helsinki, pp. 445–453

Jones, D. I. H. and Hayward, M. V. (1975) *Journal of the Science of Food and Agriculture,* **26**, 711–718

Mehrez, A. Z. and Orskov, E. R. (1977) *Journal of Agricultural Science, Cambridge,* **88**, 645–650

Morrison, I. M. (1972) *Journal of the Science of Food and Agriculture,* **23**, 455–463

Norris, K. H., Barnes, R. F., Moore, J. E. and Shenk, J. S. (1976) *Journal of Animal Science,* **43**, 889–897

The effect of different washing procedures on the losses of OM and N from samples of hay incubated in nylon bags within the rumen of sheep

M. S. Kridis[1], N. W. Offer[1] and G. D. Barber[1]

A limitation of the nylon bag technique is poor between-bag reproducibility. A major source of variation may be inconsistent post-incubation washing. Chenost *et al.* (1970) showed that additional washing in acid-pepsin reduced between-bag variability and improved prediction of forage digestibility. Recently, automatic washing machines have been used to standardize the washing procedure. We have investigated the possibility of further reducing bag variability by combining machine washing with a range of other procedures.

Method

Approximately 5 g samples of chopped (5 mm), sieved (45 μm) hay, obtained from a well-mixed bulk were incubated for 3, 8, 24 and 72 h in the rumens of four

1 Nutrition & Microbiology Department, West of Scotland College, Auchincruive, Ayr, UK.

fistulated wether sheep fed a constant high forage diet (900 g/day meadow hay and 200 g/day Ewebol sheep concentrate). Three washing agents were tested: neutral detergent reagent (ND, Van Soest and Wine, 1967), acid-pepsin (AP, Tilley and Terry, 1963) and Persil non-biological automatic washing powder (WP). ND and AP treatments were carried out by soaking bags in the reagent contained in 5 litre beakers. WP was added to the washing machine (60 g per wash) in the normal domestic manner using a hot wash (95°C), 'heavy soil' cycle. Washes were applied before – (NDI, API and WPI); after – (IND, IAP, IWP) or before and after incubation (I) in the sheep (NDIND, APIAP and WPIWP). All bags received a final cold water wash in the washing machine; control bags (C) received only this. The bags were dried at 60°C and then weighed.

Results and discussion

The model of Orskov and McDonald (1979) was applied to losses of organic matter (OM) and nitrogen (N) from the bags (Table 1). ND, AP and WP increased OM losses from bags although the forms of the degradation curves were unaltered. Washing treatments reduced between-bag variability for OM loss. OM disappearance was greatest when washing reagents were used both before and after incubation but, for single wash treaetments, it made little difference whether the wash was applied before or after incubation. Wash treatments greatly increased N losses from bags. ND and WP did not however alter the form of the disappearance curve, but for bags receiving the AP treatment, N loss was unaffected by period of incubation and so no curve was fitted.

Table 1 Effects of different washing procedures on rumen degradation characteristics

Parameter	Washing procedure*									
	C	NDI	API	WPI	IND	IAP	IWP	NDIND	APIAP	WPIWP
EDOM	0.454	0.498	0.573	0.545	0.520	0.520	0.550	0.555	0.588	0.595
CVOM24	8.0	4.3	2.2	4.3	1.8	5.2	4.0	7.6	2.0	2.4
EDN	0.538	0.648	–	0.733	0.723	–	0.790	0.778	–	0.843
N72	7.4	6.8	8.3	6.3	4.3	5.3	3.3	3.9	5.7	2.8

EDOM, EDN: effective degradability of OM and N respectively at $k = 0.04$.
CVOM24: coefficient of variation (%) for 24-h OM disappearance.
N72: concentration of N(g/kg DM) in bag residues after 72 h.

* For washing procedure–see text.

Detergent washing (ND and WP) improved bag reproducibility. Absolute losses of OM and N were increased but the unaltered forms of the degradation curves suggest that the results are as biologically meaningful as those obtained with a cold water wash only. Detergent treatments may remove contaminating residues both from the bag material and contents. Lower N contents of residues following 72-h incubation suggest that detergent washing may remove adhering bacteria from bag contents. Mathers and Aitchison (1981) have suggested this as a cause of under-estimation of protein degradability by the bag method.

Conclusion

Post-incubation detergent washing of bags appears a useful addition to the bag technique, reducing between-bag variability. Further work is required to measure the precise effect on contamination of bag residues and to assess the biological significance of the procedure. Use of domestic washing powder in the washing machine is both cheap and convenient.

References

Chenost, M., Grenet, E., Demarquilly, C. and Jarrige, R. (1970) *Proceedings of the XI International Grassland Congress,* Queensland, pp. 697–701

Mathers, J. C. and Aitchison, E. M. (1981) *Journal of Agricultural Science, Cambridge,* **96**, 691–693

Orskov, E. R. and McDonald, I. (1979) *Journal of Agricultural Science, Cambridge,* **92**, 499–503

Tilley, J. M. A. and Terry, R. A. (1963) *Journal of the British Grassland Society,* **18**, 104–111

Van Soest, P. J. and Wine, R. H. (1967) *Journal of the Association of Official Analytical Chemists,* **50**, 50–55

Acid detergent insoluble nitrogen (ADIN) as a measure of unavailable nitrogen for ruminants

C. J. Waters[1]

ADIN in forages has long been employed as a measure of non-enzymatic browning or heat damage (Van Soest, 1965), the ADIN being considered as both undegradable and indigestible (Pichard and Van Soest, 1977). This concept has been extended to include non forage feeds and has been included in new protein evaluation systems for ruminants (Webster, 1987). Recent work has questioned this use of ADIN (Britton *et al.,* 1986) particularly in feeds containing distillery byproducts (Klopfenstein and Britton, 1987).

Method

Thirty raw materials, of plant origin, were incorporated into 37 diets which were evaluated in 10 conventional digestibility trials using wether sheep fed at about maintenance. Trials were principally designed to determine the metabolizable energy (ME) values of a range of raw materials, thus materials were not selected by ADIN content. Each diet was fed to three sheep for 24 days. Feed input and faecal output were measured during the last 10 days of each trial.

Results

Diets were divided into three groups according to their apparent digestibility coefficient (APD) of ADIN: those containing tannins (T) ($n=7$); or distillery byproducts (D) ($n=5$); and the remainder (N) ($n=25$). Mean APDs of ADIN were 0.03, 0.63 and −0.68 ($P<0.001$) with N, D and T diets respectively. The mean APD of ADIN in N diets was not significantly different from zero. The APDs of ADIN in D diets agree with other published work (Britton *et al.,* 1986). The

[1] Department of Animal Husbandry, University of Bristol, UK

negative APDs of ADIN in T diets were attributed to the formation of indigestible condensed tannin protein complexes in the gastrointestinal tract (McLeod, 1974).

Apparent digestibility of total nitrogen (TN) was inversely related to the ADIN content of the diet expressed as a proportion of TN (ADIN/TN). APD TN = 0.79 − 1.56 ADIN/TN, ($P<0.001$), in all diets. Processing and heating increase the ADIN content of feeds which has been associated with reduced nitrogen digestibility in forages (Thomas et al., 1982), but not with distillery byproducts (Britton et al., 1986). Digestibility of some ADIN in distillery byproducts has been attributed to either partial condensation of amino groups with degraded sugar residues (Van Soest and Sniffen, 1984), or to reversible binding during preliminary stages of the Maillard reaction (Hodge, 1953). A crude attempt was made to calculate the APDs of artifact ADIN in D diets. Indigenous ADIN was assumed to be indigestible and comprise 1.5 g/kg of the DM (mean value for all N and T diets). This amount was subtracted from both intake and output figures and the APDs of the remaining ADIN calculated. APDs ranged from 0.7 to 1.0, thus most of the artifact ADIN appeared to have been digested. Such calculations must be treated with caution.

The supply of truly digestible undegraded nitrogen (DUN) was estimated according to ARC (1984) and Webster (1987) by incorporating degradation data obtained in sacco with estimates of rumen solid phase outflow rate. ARC (1984) may overestimate DUN supply by ignoring indigestible ADIN, while Webster (1987) may underestimate DUN supply in D diets by ignoring the digestibility of artifact ADIN. An alternative method for calculating DUN in processed or heat damaged materials would be to recognize two types of ADIN (Van Soest, 1965). As with forages, materials containing more than a specified amount of ADIN would be deemed to contain ADIN resulting from heat damage (Van Soest, 1965). A zero digestibility could be ascribed to indigenous ADIN and a different digestibility coefficient ascribed to artifact ADIN.

Conclusion

Indigenous ADIN is undegradable and indigestible and therefore may be used as a measure of unavailable nitrogen in ruminant feeds. Processing of feeds may result in the production of artifact ADIN part of which may be digestible. Calculation of DUN in heat damaged feeds may require ascribing different digestibilities to indigenous and artifact ADIN respectively. Apparent digestibility of TN in all feeds is inversely related to their ADIN contents.

References

Agricultural Research Council (1984) *The Nutrient Requirements of Farm Ruminants.* Supplement No. 1. Commonwealth Agricultural Bureaux, Farnham Royal

Britton, R. A., Klopfenstein, T. J., Cleale, R., Goedeken, F. et al. (1986) *Proceedings of the Distillers Feed Conference*, **41**, 67–75

Hodge, J. E. (1953) *Agricultural and Food Chemistry*, **1**, 928–943

Klopfenstein, T. and Britton, R. (1987) *Proceedings of the Distillers Feed Conference*, **42**, 84–86

McLeod, M. N. (1974) *Nutrition Abstracts and Reviews*, **44**, 803–815

Pichard, G. and Van Soest, P. J. (1977) *Proceedings of the Cornell Nutrition Conference*, pp. 91–98

Thomas, J. W., Yu, Y., Middleton, T. and Stallings, C. (1982) In *Proceedings of an International Symposium*, (ed. F. N. Owens). 19–21 November, 1980, Oklahoma State University, pp. 81–98

Van Soest, P. J. (1965) *Association of Official Analytical Chemists*, **48**, 785–790

Van Soest, P. J. and Sniffen, C. J. (1984) *Proceedings of the Distillers Feed Conference,* **39**, 73–82
Webster, A. J. F. (1987) In *Feed Evaluation and Protein Requirement Systems for Ruminants,* (ed. R. Jarrige and G. Alderman). Commission of European Communities, Brussels, pp. 47–53

Technical requirements for the daily assessment of individual cows' dry matter intake in the dairy herd

O. Kroll[1], E. Maltz[2], S. Denvir[2], S. L. Spahr[3] and U. M. Peiper[2]

The dry matter intake (DMI) of individual dairy cows in a herd is a parameter that cannot be routinely measured at present. Substantial research has been carried out to provide formulae for its estimation according to a cow's performance. As long as cows are fed as a group by complete diet, DMI is considered as a mean value for the average cow in the herd. The herdsman controls the diet composition and allows the cow to consume it *ad libitum*. In this case the mean values of production, weight, DMI etc. are considered sufficient for nutritional and economic decisions.

Recently, technological development and the use of controllers and computers in the industry, have permitted continuous performance data collection and automatic supplementation of the cow with concentrates accordingly. However, compound is only one fraction of the diet. Its consumption strongly affects the intake of the other fractions that must still be given to the group as a whole. The difficulty is that accurate rationing of concentrates in the individual concentrate supplementation system (ICS), without a good estimate for the DMI of other available feeds, limits our ability to exploit efficiently the available technologies. Estimating the intake of the other available feeds becomes even more complicated when there is a large variety of agricultural wastes and recycled feedstuffs.

Method

In order to provide a working tool for an accurate ICS feeding system, the necessary data were applied to calculate the DMI of the individual cows by using the available formulae to estimate the DMI of the whole group. To examine this system, the calculated values were compared to measured ones in three experimental groups (about 60 cows in each group). Two groups were given complete diets of 50:50 and 27:73 forage:concentrates ratios *ad libitum*. The third group (ICS) was given a 50:50 diet supplemented individually with concentrates through automatic self feeders for cows producing more than 30 kg milk daily (Maltz, Kroll and Spahr, 1987).

Daily milk yield was recorded from individual identification (ID) and electronic milk meters, daily bodyweight from electronic walk-through scales (located on the route out of the milking parlour) and a gate ID system, both connected to a computer, and time post-partum from the herd management program.

The complete diets were fed to the three treatment groups using a weighing mixer wagon with a controller for accurate recording of feedstuff weights.

The measurement periods were for 6 weeks in the ICS and 27:73 forage:concentrate ratio groups, and 5 weeks in the 50:50 groups. Two formulae

[1] Hahaklait Veterinary Service, Israel Cattle Breeders' Association, Tel Aviv, Israel.
[2] Institute of Agricultural Engineering, ARO Volcani Centre, Bet Dugan, Israel.
[3] Department of Animal Science, University of Illinois, Urbana, USA

(Vadiveloo and Holmes, 1979; Mertence, 1985) were tested for accuracy of DMI prediction. Intakes, both measured and calculated were determined daily and standard errors (SEs) calculated for the treatment mean using days as replicates.

Results and discussion

The two formulae used correlated differently with the actual DMI measured in both groups fed complete diets. There is an indication that formula 1 (Mertence, 1985) correlated better with the 50% forage diet group, and formula 2 (Vadiveloo and Holmes, 1979) with the 27% forage diet group (Table 1). In the ICS group the two formulae demonstrated a similar correlation with the measured DMI (Table 1).

Table 1 Actual mean daily DMI and calculated mean DMI according to performance in cows given different diets

Week	n	Actual DMI (kg/cow per day)	Calculated DMI formula 1 (kg/cow per day)	Calculated DMI formula 2 (kg/cow per day)
27:73 forage:concentrates ratio				
1	60	21.97 ± 1.67	20.74 ± 3.98	18.24 ± 2.88
2	64	19.51 ± 1.69	19.33 ± 3.53	20.18 ± 2.69
3	65	22.79 ± 1.31	19.41 ± 3.01	21.80 ± 2.91
4	68	21.92 ± 1.31	19.76 ± 3.00	21.78 ± 2.80
5	66	22.87 ± 1.56	19.80 ± 3.14	17.83 ± 2.47
6	69	22.68 ± 1.15	20.76 ± 3.57	21.53 ± 3.02
50:50 forage:concentrates ratio				
1	55	20.94 ± 2.03	20.77 ± 3.82	17.72 ± 2.46
2	58	18.90 ± 2.01	19.87 ± 3.01	17.12 ± 2.14
3	58	19.55 ± 0.87	19.06 ± 3.65	17.81 ± 2.12
4	61	20.57 ± 1.79	18.75 ± 3.64	17.81 ± 2.14
5	65	20.97 ± 1.05	20.25 ± 5.10	17.83 ± 1.93
ICS				
1	53	21.31 ± 1.89	21.66 ± 4.04	18.48 ± 3.54
2	58	19.66 ± 1.90	20.16 ± 3.54	18.31 ± 3.83
3	56	21.91 ± 1.02	19.18 ± 3.54	19.87 ± 4.14
4	61	21.39 ± 0.21	19.48 ± 3.68	19.74 ± 4.04
5	61	22.39 ± 1.93	19.33 ± 3.46	19.74 ± 3.93
6	60	23.04 ± 1.18	19.97 ± 5.33	19.86 ± 4.03

Formula 1 – Mertence, 1985.
Formula 2 – Vadiveloo and Holmes, 1979.
DMI – dry matter intake.

Analysing the two formulae on individual cow's data (Table 2) suggests that the higher the supplementation level the greater the diversion between the two formulae. The individual estimates according to the two formulae may differ largely (Table 2), but it is not reflected in the mean values (Table 1). Each formula may be applicable in a different complete diet system, but the temptation to apply either of them to individual cows in an ICS system should be avoided. It is necessary to examine each formula according to the herd feeding regimen. In our

Table 2 A sample of individual bank DMI of cows in the ICS group given 50:50 forage:concentrate ratio, and the parameter used to calculate it. Individual supplementation for cows producing over 30 kg milk per day

Cow no.	Week post-partum	Mean weekly bodyweight (kg)	Mean bodyweight change during the week (kg/day)	Milk yield (kg/day)	Butter fat (%)	Concentrates supplementation DM (kg/day)	Calculated DMI	
							Formula 1	Formula 2
54	21	556	1.60	28.3	3.1	–	21.59	17.11
57	27	570	-0.24	26.0	3.8	–	16.32	16.61
105	14	557	0.24	30.9	4.1	2.7	17.04	16.54
642	22	534	0.96	32.0	3.8	2.7	18.96	16.43
401	19	659	0.88	37.8	3.1	5.4	17.05	19.17
119	16	509	-0.96	37.2	3.9	6.3	11.06	15.72
1119	13	730	0.20	46.0	3.3	8.1	15.97	21.36
1898	7	661	-0.92	44.1	2.9	8.1	10.25	18.88
634	7	647	0.16	48.9	3.1	9.0	14.69	19.48

DMI – dry matter intake.

case, it seems that formula 1 is applicable to the ICS group with a correction of +10–15% to the highest supplemented cows and a gradual reduction in the correction factor to zero when no supplementation is given.

References

Maltz, E., Kroll, O. and Spahr, S. L. (1987) *Automation in Dairying 3rd Symposium*, Wageningen, The Netherlands, September 9–11, 1987

Mertence, D. R. (1985) *Georgia Nutritional Conference*, Atlanta, Georgia. February 13–15, 1985

Vadiveloo, J. and Holmes, W. (1979) *Journal of Agricultural Science, Cambridge*, **93**, 553–562

A technique to record eating and ruminating behaviour in dairy cows

C. A. Huckle[1], A. J. Clements[1] and P. D. Penning[2]

The way in which cows behave under various conditions will have a direct effect on their feed intake and hence, ultimately, on their level of milk production. Previously, behavioural studies on dairy cows have been time consuming, laborious exercises impractical in many situations and difficult to carry out continuously. Equipment which enables recordings of ingestive behaviour of sheep and beef cattle has now been developed for use on dairy cows permitting the automated recording of behaviour for a 24-h period (Figure 1). It has been designed to cause

Equipment

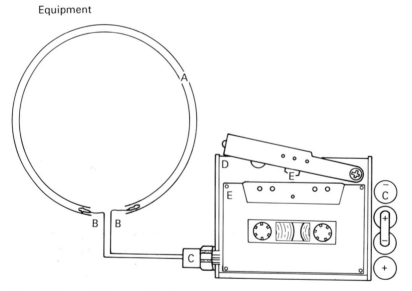

Figure 1 Schematic diagram of principal components of behaviour recorder. A: Transducer (silicon tubing packed with graphite and morganite); B: electrodes; C: connection to recorder via capacitor and resistor; D: MR10 miniature tape recorder; E: audio cassette C120; F: recording head; G: battery pack

[1] AFRC Institute for Grassland and Animal Production, North Wyke Research Station, North Wyke, UK
[2] AFRC Institute for Grassland and Animal Production, Hurley, UK

minimal disturbance to the cow and to withstand conditions associated with the grazing animal.

Equipment

A resistive transducer (A) is fitted around the jaw of the animal in the form of a noseband. The electrodes (B) are embedded in the graphite and morganite conductor and relay the signal to an Oxford MR10 miniature tape recorder (D) via a capacitor and resistor (C). Both units are mounted on a head collar worn by the animal. When a voltage from the recorder is applied to the transducer, the jaw movements of the animal cause stretching of the transducer which in turn causes a change in resistance, creating analogue signals, which are converted to frequencies and stored on cassette by the tape recorder.

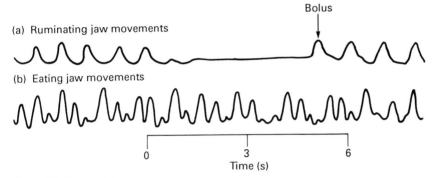

Figure 2 Oscillograph trace

The recorded signals are subsequently replayed at 60 times the recording speed on an Oxford PB2 replay unit. The signal is amplified by an Oxford PM3 amplifier, this amplifier has two outputs for the signal, one to a 20 MHz oscilloscope or an ultraviolet oscillograph, the other to a microprocessor, which carries out the analysis. Detailed summaries of the time spent eating, ruminating and idling together with information on rates of jaw movements during eating and ruminating are produced (Figure 2).

Conclusions

(1) Causes minimal disturbance to the cow and will withstand harsh conditions.
(2) Enables observations for a full 24-h period.
(3) Enables a full range of events to be recorded.
(4) Allows several observations to be taken simultaneously.

Reproductive technology

Twinning cattle by embryo transfer

P. J. Broadbent[1] and D. F. Dolman[1]

The rate of reproduction in cattle is a limitation to their biological and economic efficiency, particularly in the case of suckler cows. Induction of twinning in cattle would improve reproductive efficiency but the economic consequences depend on the costs associated with doing so as well as the potential increase in output.

Method

The most reliable method of inducing twin calf production is by embryo transfer. This technique has been studied with 358 cows and maiden heifers, mainly Hereford × British Friesians (HFr), over a 4-year period. These potential recipients had their oestrous cycles synchronized and they were all mated using Charolais semen to establish an indigenous embryo.

Superovulated HFr cows were mated using Simmental semen and flushed non-surgically 7 days after oestrus to provide embryos which were used fresh or frozen/thawed for transfer to a recipient.

Embryos were transferred only to those recipients with a corpus luteum detected by palpation *per rectum*. Most transfers were made via flank incision, with the recipient suitably sedated and anaesthetized, but some have been made non-surgically via the cervix. The transferred embryo was placed in the uterine horn contralateral to the corpus luteum irrespective of transfer technique. Each animal was subjected to these procedures once and subsequently mated by natural service.

Results

Embryos were transferred to 311 of the 358 potential recipients (86.8%) and the combined artificial insemination (AI)/embryo transfer procedure resulted in pregnancies in 199 cows and heifers with 89 multiple pregnancies being created. These figures represent a pregnancy rate of 55.6% and a multiple pregnancy rate of 24.8% of all potential recipients.

Expressed in relation to those recipients receiving embryos by transfer, the pregnancy rate was 64% and the multiple pregnancy rate 28.6%. Single pregnancies consequent on the survival of the transferred embryo in the absence of the indigenous embryo have occurred at the rate of 4–12% of all pregnancies. In the latest group of animals for which calving data are available there were 41 potential recipients and 40 of these received a transferred embryo. There were 26 animals pregnant to the single, combined AI/embryo transfer procedure which represents 63.4% and 65% of potential and actual recipients respectively. The 16 multiple pregnancies represented 39% and 40% of potential and actual recipients respectively, and produced 28 live calves.

[1] North of Scotland College of Agriculture, 581 King Street, Aberdeen, UK

Within this group of animals there were 15 well grown maiden heifers which had an overall pregnancy rate of 73.3% with a multiple pregnancy rate of 53.3%. The eight twin-bearing heifers produced 15 live calves.

Discussion

Twinning rate is the product of the pregnancy rate that could be achieved by AI or natural service in the recipients and the survival rate of the transferred embryos. Twinning rate has increased with improvements made to embryo transfer techniques during the course of this work. When good transfer techniques were combined with high normal pregnancy rates then a twinning rate of over 50% was achieved by one exposure of each cow to the combined AI/embryo transfer technique which is consistent with the data of Diskin, McDonagh and Sreenan (1987). If cows that returned to oestrus after their first exposure to the technique were recycled and given further exposures then the overall twinning rate could be raised to 45–70% of all cows. Moreover, of single pregnancies in cows given transfers, an appreciable proportion are found to result from the transferred embryo.

Using embryos produced by superovulation for twinning commercial beef and dairy cows would be uneconomic. However, recent advances in the production of bovine embryos by technology *in vitro* (Gordon *et al.*, 1988) will enable embryos to be produced relatively cheaply. Twin calf production will involve additional labour, feed, managerial and veterinary inputs as well as different skills relative to single calf production. Its application should be restricted to fertile, well managed cows to provide the farmer with a means of maintaining or improving his net income from his breeding cattle.

References

Diskin, M. G., McDonagh, T. and Sreenan, J. M. (1987) *Theriogenology,* **27**, 224 (abstract)
Gordon, I., Lu, K. H., Gallagher, M. and McGovern, H. (1988) *Animal Production,* **46**, 498 (abstract)

Effect of suckling on embryo quality from superovulated Welsh black cows

C. M. Brown, J. B. Thomas, G. Williams, J. M. Owen,
R. F. E. Axford, I. B. H. Wilson and J. B. Owen[1]

In 1985 an investigation was begun into the application of multiple ovulation and embryo transfer (MOET) techniques to genetic improvement in the Welsh Black Cattle Group Breeding Scheme, University of Wales, Bangor. It was proposed to utilize MOET methods to facilitate the transfer of genetic material between the nucleus and member herds and anticipated that by this route a degree of referencing could be achieved between the herds of animals. However, the results of 2 years of applying MOET to Welsh Black cows have revealed marked seasonal variation in the quality of embryos recovered. It was considered that suckling may have a role in this observation and an experiment was designed to test this theory.

[1] University of Wales, Bangor, UK.

Method

Twelve multiparous Welsh Black cows were randomly allocated, in pairs, to two treatment groups; weaning at 2 days post-calving or normal suckling. At a mean of 60 days post-calving the cattle were superovulated. The protocol was as follows:

(1) Progesterone releasing device (PRID) insertion for 12 days to synchronize the reference oestrus
(2) a single injection of 2500 iu pregnant mare's serum gonadotrophin (PMSG) i/m at day 10 post-oestrus (oestrus = day 0)
(3) a double dose of prostaglandin (PG) $F_{2\alpha}$ (Estrumate, Coopers Ltd., 4 ml) 48 h later
(4) insemination; two straws 12 h after the onset of oestrus followed by a single straw 12 h later.

Non-surgical recovery of embryos was performed 7 days after insemination. The number of corpora lutea (CLs) in each ovary was estimated manually by rectal palpation. Embryos recovered were graded morphologically and the significance of treatment differences examined by analysis of variance of angularly-transformed results.

Results

Overall embryo recovery was 57.6% with no significant difference between groups. Mean numbers of CLs and good and poor embryos found are recorded in Table 1.

Embryo quality, calculated as the percentage of those recovered judged to be of usable quality, was significantly better for the weaned group of cows.

Table 1 Mean response of cows to superovulation and mean quality of embryos recovered

	CLs detected	Good	Poor	Good (%)
Weaned cows	8.0	3.8	1.3	74.5
Suckling cows	10.7	1.8	3.7	32.7
Significance	NS	NS	NS	$P = 0.04$

NS: not significant

Conclusions

These data show that suckling had a detrimental effect on embryo quality in the superovulated cow. The reasons for this are unclear and further investigations will be needed to confirm the effect and identify causal factors. A possible interaction with MOET techniques could reduce the applicability of these to beef animal improvement. Specifically the genetic benefits of such schemes will need reappraisal if it is necessary to wean cows early and/or maintain them empty.

Performance and behaviour of four sex conditions in male suckled calves

C. A. Moore[1], W. McLauchlan[2], M. J. Doherty[2], W. J. McCaughey[3] and B. W. Moss[4]

The recent ban on the use of hormonal growth promoters in beef cattle in the European Community has encouraged a greater interest in the use of entire male animals for beef production. In the UK the production of beef from bulls has not been widely practised due mainly to behavioural problems.

Successful immunological castration of young bulls has been achieved through the neutralization of luteinizing hormone release hormone (LHRH) by active immunization (Robertson *et al.*, 1982). This experiment was designed to examine the performance, behaviour, fertility and carcass quality of castrated, entire, vasectomized and immunized male animals over 3 years. The results presented here refer only to the first year of the experiment.

Method

Spring-born Continental cross-bred calves were either double suckled (64 calves) or single suckled (29 calves). Castration was carried out at approximately 8 weeks and vasectomy at 10–13 weeks. The first immunization was given at 13–15 weeks of age followed by a second 7–8 weeks later.

Results

The animal performance and carcass data are summarized in Table 1.

The treatments had no significant effect on liveweight gain (LWG) during the grazing season. From housing at 7–8 months of age until slaughter at 15 months

Table 1 Performance, carcass and behaviour data

	Castrate	Entire	Vasectomized	Immunized	SED	Significance
Liveweight gain from 2 months to 15 months (kg)	0.91	0.98	1.02	1.10	0.025	*
Carcass weight (kg)	255	287	292	308	6.6	*
Cannon bone length (cm)	22.5	21.9	21.9	22.2	0.02	***
Frequency of head contact with and without retaliation	1.6	5.8	6.0	3.5	1.75	NS
Frequency of homosexual mounting	0.8	1.6	1.6	0.9	0.14	*
Frequency of mounting heifer on heat	4.4	12.1	11.2	11.5	2.66	NS

NS: not significant; * $P < 0.05$; *** $P < 0.001$.

[1] Greenmount College of Agriculture and Horticulture, Antrim, UK
[2] Castlearchdale EHF, Irvinestown, Co Fermanagh
[3] Veterinary Research Laboratories, Stoney Road, Stormont, Belfast, UK
[4] Food and Agricultural Research Division, Newforge Lane, Belfast, UK

bulls gained weight significantly faster than castrates. Within the three bull types the LWG of immunized bulls was greater than vasectomized bulls which was greater than entire bulls ($P<0.01$). A similar trend in LWG was observed over the whole period of the experiment. Silage dry matter (DM) intake was 20% higher in bulls compared with castrates, resulting in the bull types requiring approximately 1 tonne more of fresh silage per head than the castrates. The immunized bulls had the highest killing-out percentage and produced significantly ($P<0.05$) heavier carcasses than the other male types. Entire and vasectomized bulls were significantly leaner than castrates ($P<0.05$), immunized bulls had fat levels between the castrates and the other two bull groups. The length of cannon bone in immunized bulls was mid-way between the relatively long cannon bone in castrates and the shorter cannon bone in the entire and vasectomized bulls.

Aggressive and homosexual behaviour was assessed by counting the number of head contacts and homosexual mounts per animal during each observation period respectively. Entire and vasectomized bulls showed the greatest level of aggression and castrates the least. The aggression of the immunized bulls was mid-way between those two extremes although the differences were not significant. The homosexual behaviour of the immunized bulls was similar to the castrates and significantly less than the other two bull groups ($P<0.05$). Sexual activity was assessed by counting the number of mounts on heifers in standing heat. All three bull types were more sexually active than the castrates, although the differences were not significant.

The percentage binding LHRH indicated that immunization was effective after the second injection and lasted on average 3.2 months (range 0–9 months). However, 21% of the immunized bulls did not respond to treatment.

Conclusion

The initial results indicate that active immunization against the production of LHRH may be effective in reducing general behavioural problems associated with bulls, without impairing animal performance.

Reference

Robertson, I. S., Frazer, H. M., Innes, G. M. and Jones, A. S. (1982) *Veterinary Record,* 111, 529–531

Breeding

A comparison of indigenous with exotic cross-bred cattle in the dry zone of India

J. P. Mittal[1] and S. Prasad[2]

A hostile environment with temperatures ranging from 2°C in winter to 50°C in summer, coupled with a severe scarcity of fodder owing to successive droughts, lack of drinking water and saline nature of underground water, are some of the

[1] Division of Animal Studies, Central Arid Zone Research Institute, Jodhpur, India
[2] Regional Research Station, Central Arid Zone Research Institute, Bikaner, India

constraints to livestock development in the dry zone in India. Looking towards the farmers of other parts of the country, some of the farmers in this area have started crossing their cattle with improved European breeds such as Holstein, Jersey and Red Dane. This practice is more common in cities. This report provides some facts, based on surveys, regarding the suitability of European cattle breeds for improving the local stock of the dry zone of India.

Method

Regular detailed surveys of various cities and representative villages of the hot arid zone of India were conducted by a team of animal production scientists between 1982 and 1987. Data and information were collected on type of cattle, their feeding practices, productive performance, and general health.

Results

Type of cattle

The main indigenous breeds found were Tharparpar, Gir and Rathi (dairy), Kankrej and Haryana (dual purpose), Nagauri (draught); exotic cross-bred cattle were Holstein, Jersey and Red Dane crosses.

Feeding practices

While seasonal grazing was practised in native cows, the cross-bred cows were not grazed which added to their maintenance cost. Generally cows were fed *ad libitum* roughages (wheatstraw), pala leaves (*Ziziphus numularia*) and chaffed *Lasiurus sindicuss* grass, and the cross-bred cows consumed 8–12 kg roughage daily while the daily consumption was 5–8 kg in native cows. The cross-bred cows were provided with green fodder in urban areas at almost all times of the year which was not the case with native cows. Concentrate mixture, consisting of wheatbran, gram husk, mung husk and cottonseed cake, was also provided to each productive cow. The cross-bred cows were offered 6–10 kg of this mixture daily while native cows were provided with only 3–5 kg.

Productive performance

Well fed native cows produced their first calf at the age of 3 years in the urban sector whereas most of the cross-bred cows calved before 3 years. Native cows in the urban areas had a longer dry period (3–6 months) than all types of crosses (2–4 months). In rural areas the intercalving period in cross-bred cows was 12–14 months and in native cows 14–18 months.

In the urban sector, cross-bred cows were superior in their milk production to native cows (Holstein cross 12.1 litres/day, Red Dane cross 10.2 litres/day, Jersey cross 9.7 litres/day, Rathi 9.5 litres/day, Gir 8.1 litres/day and Tharparpar 7.3 litres/day). In the rural sector native cows yielded more milk than cross-bred cows irrespective of their types. Although milk yield was higher in cross-bred than native cows in urban areas the cost of milk production per litre was almost the same in both cases, whereas the risk in keeping native cows was less (Prasad and Mittal, 1986, 1987). Cross-bred cows had a consistent reduction in their milk yield (50–60% in urban and 70–80% in rural sectors) during summer, whereas in native cows it was virtually nil in the cities and was marginal (5–10%) in villages.

General health

The adverse climate and various constraints had very little or no effect on the health of native cattle in any season, whereas all types of crosses were seen panting with open mouths and salivating during the daytime in the summer months. In some cases skin eruptions and excessive lacrimal discharge were noticed.

Calf survival of cross-breds was lower (40–50%) than native cattle (80–90%). The incidence of mastitis was high in cross-bred cows and they were more prone to suffer from parasites. In addition, if sick cross-bred animals were not treated within hours of disease onset it was often fatal, which was not the case with native animals.

References

Prasad, S. and Mittal, J. P. (1986) *National Seminar on Cross-bred Technology in Increasing Milk Production.* JNKVV, Jabalpur-India, November 17–19

Prasad, S. and Mittal, J. P. (1987) *Indian Farming,* **37**, 27–28

Rathi – a new breed of cattle from the Indian desert

J. P. Mittal[1] and S. Prasad[2]

Most cattle breeds in the North-West region of India are heavy and large (Kaura, 1955). Due to resource constraints in this region, a breed having a smaller body size, earlier maturity and higher milk yield/kg liveweight would be advantageous. A newly discovered breed of cattle from this region, the Rathi, (Figure 1) possesses all these traits (Mittal and Prasad, 1988).

Figure 1 A cow of the Rathi breed

[1] Division of Animal Studies, Central Arid Zone Research Institute, Jodhpur, India
[2] Regional Research Station, Central Arid Zone Research Institute, Bikaner, India

Area of availability

Rathi cattle are found in Bikaner Ganganagar and Jaisalmer districts of arid North-West India (27–30°N; 72–75°E). Being in the heart of the Thar desert the environment for these cattle is very hostile. The present population is about 500 000 head. Bodyweights and measurements of Rathi heifers at different stages of growth are provided in Table 1.

Table 1 Liveweight and allometric measurement of Rathi heifers

	Liveweight (kg)	Heart girth (cm)	Pin-shoulder length (cm)	Height at withers (cm)
At birth	19.5 ± 0.35	61.0 ± 0.95	55.5 ± 1.09	68.0 ± 1.21
At 1 month	35.9 ± 0.86	73.5 ± 1.02	67.7 ± 0.92	72.1 ± 1.42
At first conception	227.8 ± 3.12	130.1 ± 1.98	123.0 ± 0.85	118.9 ± 2.04
At first calving	247.7 ± 2.94	150.2 ± 2.14	130.5 ± 1.36	123.2 ± 1.88
At adulthood	295.0 ± 4.01	162.8 ± 3.20	135.4 ± 1.84	125.0 ± 1.75

Adaptational characteristics

(1) Smaller size: feed requirements are comparatively lower
(2) Short and smooth body coat: indicative of lower metabolic rate and capacity to react favourably to stress
(3) Heat tolerant: heat tolerance coefficient of free grazing animals is very high (89.87)
(4) Thermostable: average fluctuation from morning to evening in physiological responses are:

rectal temperature $0.503 \pm 0.021°F$
respiration rate $6.17 \pm 0.143/min$
pulse rate $15.52 \pm 0.236/min$

(5) Low body water turnover rate: 281 ml/kg bodyweight 0.82/24 h
(6) Early maturity: age at maturity is the lowest of all Indian breeds
(7) Short dry and intercalving period.

Satisfactory production has been recorded in a regional survey (Tables 2 and 3) even in the extreme conditions under which this breed is kept.

Table 2 Production characteristics of Rathi cows

Parameters	Mean + s.e.	Range
Age at maturity (months)	22.5 ± 0.52	18–28
Age at first calving (months)	36.30 ± 0.84	27–40
Daily milk yield (litres)	9.52 ± 0.25	4–16
Lactational yield (litres)	2810 ± 6.04	2200–3300
Lactational length (days)	310.3 ± 3.42	280–330
Dry period (days)	165.5 ± 2.13	150–185
Calving interval (days)	492.8 ± 4.15	420–540

Table 3 Milk composition in Rathi cows

Attributes	Mean	s.e.
Specific gravity	1.03	0.04
Total solids (%)	12.55	0.32
Fat (%)	3.68	0.12
Lactose (%)	4.83	0.05
Protein (%)	3.35	0.14
Ash (%)	0.86	0.01

References

Kaura, R. L. (1955) *Indian Breeds of Livestock*. Prem Publisher, Golanganj, Lucknow
Mittal, J. P. and Prasad, S. (1988) *Proceedings of the VI World Conference of Animal Production*. Helsinki, 4.122

Computer studies

A natural-language interface for retrieval of information from a dairy database

L. R. Jones and S. L. Spahr[1]

A natural-language interface (NLI) was developed to:

(1) provide flexible access to dairy herd records stored in an on-farm microcomputer database
(2) provide knowledge based replies to questions concerning reproductive status, daily yield and identification of individual animals or groups of animals.

The NLI was developed in the LISP artificial intelligence language using GCLisp (Gold Hill Computers, Cambridge, MA) to allow LISP programming on an IBM-AT-compatible microcomputer, and dBLisp (Chestnut Software, Cambridge, MA) to access of dBASE III files from Dairybase, the University of Illinois on-farm database used for development.

Operation of the program was initiated by a user entering a natural-language question or command-type (e.g. List ---) sentence about the database. A full screen editor assisted the user in correcting mis-spelled words by suggesting alternative words for unknown words. Parsing was accomplished using augmented transition network (ATN) techniques. The ATN used LISP functions as continuations to achieve chronological backtracking. Syntactic rules, which described the grammar of the input, were represented in a high-level language and converted to LISP functions using an interpreter specifically developed for the project. After parsing, the syntactic structures were converted to semantic concepts representing the underlying notions of the input using semantic rules which were implemented as LISP functions. The semantic concepts were stored in a data dictionary and

[1] Department of Animal Sciences, University of Illinois at Urbana-Champaign, Urbana, IL, USA

represented using a discrimination network. Concepts were converted to answers by the response module, using information stored with the concepts to calculate and display the answer. Conversion of the input to semantic concepts was termed 'conceptual semantics' and is believed to be a new artificial-intelligence method.

Concepts relating to dairy reproductive management were identified and incorporated into a data dictionary. Concepts corresponded to variables and functions of variables stored in the database. The use of functions of variables allowed the NLI to answer questions about information not stored explicitly in the database (e.g. due to be pregnancy checked). Associated with the concepts were definitions and procedures for printing the answer in either a knowledge-based natural language format or as a table.

Random access memory (RAM) requirements to operate the NLI depended on the amount of memory required to load the LISP environment along with the NLI and the amount of memory used for specific questions. Loading LISP and the NLI used 27% of the 5 megabytes of RAM memory available in the test system. Memory required to operate the NLI was dependent on the complexity of the input. Parsing occurred in RAM memory and typically required 4–8 s, depending on the complexity of the query. After parsing, retrieval of information from the database and formulation of the answer required from 1 s to 5 min depending on the number of animals in the request, the complexity of the request (number of database searches required) and the size of the database. Response time for questions involving a group of animals was reduced when a tabular output was used instead of a natural-language answer.

The NLI was capable of answering queries about individual animals or subgroups within a herd. Subgroups were limited only by data fields in the database. Subgroups tested included those based on breed, age, location, sire, dam, service sire, daily milk yield and various expressions of reproductive status. Answers for multiple animals could be either in a natural language form (knowledge-based, complete set of information to aid in decision making) or in a tabular form (specific, concise answer). Answers also could be counts or means of data fields within specific subgroups.

It is concluded that retrieving information from a dairy database using natural-language queries on a microcomputer is a feasible alternative to traditional approaches of information retrieval. It was shown that by using the 'conceptual semantics' approach to answer formulation, the NLI method could provide a complete answer with additional pertinent information. The resulting NLI developed in this study demonstrated that NLI answers could provide the user with information concerning a variety of situations related to reproductive management of dairy cattle that were superior to answers obtained from traditional dairy-herd-management systems.

Energy partitioning in cattle by a microcomputer model: a teaching aid

B. Walker, B. A. Young and G. Godby[1]

Practical animal feeding systems are based on estimated energy requirements for maintenance and production. The equations used to calculate these requirements

[1] Department of Animal Science, University of Alberta, Edmonton, Alberta, Canada

are derived from the extrapolation of a large number of experiments. Under practical conditions, animal performance often deviates from predicted values as a consequence of the interplay of factors not accounted for by the simplistic, factorial method of estimating requirements. Computer modelling provides a possible mechanism to overcome these problems.

Method

Partitioning of daily energy intake into waste products, heat and the products of growth and reproduction has been represented in a funnel diagram (Young, 1975). The model directs energy preferentially to maintenance then to the products of gestation and lactation, and the balance, positive or negative, to body tissues. Input to the model is at optional levels of complexity through expansion fields for environment, feed and milk production, and output is graphical and numeric. The equations used in the model were derived mainly from MAFF (1975) and NRC (1981). One of three parameters (feed intake, milk yield or liveweight change) is predicted based on inputs of the other two and energy partitioning from a single metabolizable energy (ME) pool is balanced in an iterative procedure.

The model was used to evaluate the effects of cold ambient temperatures on the relationship between energy intake and energy retention in cattle. The liveweight change of a 450 kg cow, dry and 150 days pregnant, was predicted at several levels of ME intake (60 to 130 MJ/day of a hay diet containing 9.13 MJ ME/kg) and exposed to effective ambient temperatures between 10 and −20°C. The animal was assumed to be kept in confinement, adapted to 10°C. The model was run for animals in thin, average or fat body conditions using appropriate animal thermal resistance values (0.3–0.45°C.m²/watt; Young, 1975).

Results

There was a curvilinear relationship between predicted daily liveweight change (energy retention) and feed intake over the range from 60 to 130 MJ ME/day

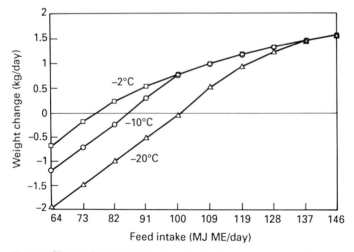

Figure 1 Changes in energy retention with feed intake and ambient temperature

(Figure 1). The lower critical temperature of the cow fed at maintenance (76 MJ ME/day) was predicted to be $-2°C$. With colder temperatures there was a progressively greater but parallel weight loss over the lower half of the curve and hence an increase in the predicted maintenance requirement (ME intake for zero energy retention). Above maintenance the lines converged with that of the animal when not cold stressed ($>-2°C$). The greater slope implies an apparently higher partial efficiency of gain under conditions of cold stress. This has been observed experimentally in pigs (Close and Mount, 1976). At temperatures below 0°C reduced body fatness increased cold susceptibility, and shifted the relationship between intake and energy retention in a similar manner as reduced ambient temperature.

Discussion

The test scenario used illustrated some of the basic, underlying principles of bioenergetics which can be communicated through the model. In addition, the practical significance of increased cold susceptibility of cattle fed at low levels during the winter months was eluded to. The model can be of assistance as an interactive teaching aid and in management decisions without the need for lengthy and costly animal experimentation.

References

Close, W. H. and Mount, L. E. (1976) *Proceedings of the Nutrition Society,* **35**, 60A

MAFF (1975) *Technical Bulletin 33.* Her Majesty's Stationery Office, London

NRC (1981) *Effect of Environment on Nutrient Requirements of Domestic Animals.* National Academy Press, Washington, DC

Young, B. A. (1975) *Brody Memorial Lecture XII*, University of Missouri – Columbia. Special Report 175

List of participants

Luigi Albertini
Agrilatte Sistemi, Via Aurelia Km 29, Torrimpietra 0050, Italy

Jack L. Albright
Department of Animal Sciences, Poultry Science Building, Purdue University, West Lafayette, Indiana, USA

Geoff Barber
West of Scotland College, Auchincruive, Ayr, UK

John A. Bax
West of Scotland College, Crichton Royal Farm, Mid Park, Bankend Rd, Dumfries, UK

Steve Bettany
Farmelectric Centre, National Agricultural Centre, Stoneleigh, Warwickshire, UK

Timothy D. A. Brigstocke
BOCM Silcock Ltd, Basing View, Basingstoke, Hampshire, UK

Peter J. Broadbent
North of Scotland College of Agriculture, Animal Husbandry Division, School of Agriculture, 581 King Street, Aberdeen, UK

Christopher Brown
School of Agricultural and Forest Sciences, University of Wales, Bangor, Gwynedd, UK

Jim Bruce
Scottish Farm Buildings Investigation Unit, Craibstone, Bucksburn, Aberdeen, UK

A. Gay Buchanan
Elanco Products Ltd, Dextra Court, Chapel Hill, Basingstoke, Hampshire, UK

Anthony G. Chamberlain
School of Agricultural and Forest Sciences, University of Wales, Bangor, Gwynedd, UK

A. June Clements
Institute for Grassland and Animal Production, North Wyke, Okehampton, Devon, UK

Michael Coffey
British Friesian Cattle Society, Scotsbridge House, Rickmansworth, Hertfordshire, UK

Ian R. Cumming
Somerset Cattle Breeding Centre, Ilminster, Somerset, UK

Robert W. Davies
Farmers' Weekly, 4A High Street, Welshpool, Powys, UK

Chris Davis
Elanco Products Ltd, Dextra Court, Chapel Hill, Basingstoke, Hampshire, UK

Roger Ewbank
University Federation for Animal Welfare, 8 Hamilton Close, Potters Bar,
Hertfordshire, UK

Aase Marit Flittie Anderssen
NRF – Norwegian Cattle, Vagamo, Norway

J. Michael Forbes
Department of Animal Physiology and Nutrition, Leeds University, Leeds, UK

Fraser Donald
Lopen Feed Mills, Cross Hands, Llanelli, Dyfed, UK

Phil C. Garnsworthy
School of Agriculture, University of Nottingham, Sutton Bonington, Loughborough,
Leicestershire, UK

James F. D. Greenhalgh
School of Agriculture, Aberdeen University, Aberdeen, UK

Richard D. Gill
BOCM Silcock Ltd, Basingstoke View, Basingstoke, Hampshire, UK

Gunnela M. Gustafson
Svante Gustafson
Animal Nutrition and Management, Kungsangens Gard, SLU, Uppsala, Sweden

Khaled Hecheimi
School of Agricultural and Forest Sciences, University of Wales, Bangor, Gwynedd,
UK

Christopher A. Huckle
Institute for Grassland and Animal Production, North Wyke, Okehampton, UK

Jeremy Hughes
Carrs Farm Food Ltd, Old Croft, Stanwix, Carlisle, UK

R. C. Jackmola
School of Agriculture, University of Aberdeen, 581 King Street, Aberdeen, UK

Nicola James
School of Agricultural and Forest Sciences, University of Wales, Bangor, Gwynedd,
UK

Edmund B. Jarvis
Colborn-Dawes Nutrition Ltd, Heanor Gate, Heanor, Derbyshire, UK

D. Iorweth H. Jones
Welsh Plant Breeding Station, Institute for Grassland and Animal Production, Plas
Gogerddan, Bow Street, Aberystwyth, Dyfed, UK

Michael Kelly
Farm Buildings Department, West of Scotland College, Auchincruive, Ayr, UK

John W. B. King
Edinburgh School of Agriculture, West Mains Road, Edinburgh, UK

Sarah A. Kitwood
School of Agricultural and Forest Sciences, University of Wales, Bangor, Gwynedd,
UK

Offer Kroll
Israel Cattle Breeders' Association, 25 Arlozorov Street, Tel Aviv, Israel

J. David Leaver
Wye College, University of London, Wye, Nr Ashford, Kent, UK

David G. Lloyd
Daily Post, Sans Souci, Rhewl, Gobowen, Oswestry, UK

Fayez I. Marai
Department of Animal Production, Faculty of Agriculture, Zagazig University, Zagazig, Egypt

Jagdish P. Mittal
Rowett Research Institute, Bucksburn, Aberdeen, UK

Charlotte Moore
Greenmount College of Agriculture and Horticulture, 22 Greenmount Road, Antrim, UK

Ingrid M. Mossberg
Animal Nutrition and Management, Kungsangens Gard, Uppsala, Sweden

Julian P. Murray-Evans
School of Agricultural and Forest Sciences, University of Wales, Bangor, Gwynedd, UK

Peter A. Murrell
Constance Murrell, Farming Supplies (Maldon) Ltd, Millers, Maypole Road, Wickham Bishop, Nr Witham, UK

John C. Nolan
Colborn-Dawes Nutrition Ltd, Heanor Gate, Heanor, Derbyshire, UK

Nicholas W. Offer
West of Scotland College, Auchincruive, Ayr, UK

Hussain Omed
School of Agricultural and Forest Sciences, University of Wales, Bangor, Gwynedd, UK

John B. Owen
School of Agricultural and Forest Sciences, University of Wales, Bangor, Gwynedd, UK

William J. A. Payne
Hillhay Cottage, Broadway, Hereford and Worcester, UK

Clive J. C. Phillips
School of Agricultural and Forest Sciences, University of Wales, Bangor, Gwynedd, UK

Gianfranco Piva
Facolta Agraria, Via Parmense 84, Piacenza, Italy

John H. D. Prescott
Wye College, Wye, Nr Ashford, Kent, UK

T. Reg Preston
Convenio Interinstitucional para Produccion Pecuaria en el Valle del Rio Cauca, Foundation para del Desarrolo Integral del Valle del Cauca de Camea, Calle 8a, No.3–14 Piso 17, Cali, Colombia

Jim F. Roche
Department of Animal Husbandry, Faculty of Veterinary Medicine, Ballsbridge, Dublin, Eire

Stuart Revell
Milk Marketing Board, Freezing Board, Freezing Unit, MMB, Llanrhydd, Ruthin, Clwyd, UK

David W. B. Sainsbury
Department of Clinical Veterinary Medicine, Madingley Road, Cambridge, UK

S. Anne Schofield
Volac Ltd, Orwell, Royston, Hertfordshire, UK

Patricia Simpson
Butterworth Scientific Ltd, PO Box 63, Westbury House, Bury Street, Guildford, Surrey, UK

Sidney L. Spahr
Department of Animal Sciences, University of Illinois at Urbana – Champaign, 315 Animal Science Laboratory, 1207 West Gregory Drive, Urbana, Illinois, USA

W. Ray Stricklin
Department of Animal Sciences, University of Maryland, College Park, Maryland, USA

Phil C. Thomas
West of Scotland College, Auchincruive, Ayr, UK

Fredrick Thompson
Rumenco Ltd, Stretton House, Burton-on-Trent, Staffordshire, UK

Mats Tornquist
Animal Health Service, Kammahareg 10, Skara, Sweden

H. J. Van de Merwe
Department of Animal Science, University of Orange Free State, PO Box 339, Bloemfontein, South Africa

Beth Walker
Department of Animal Science, University of Alberta, Edmonton, Alta, Canada

R. C. Waters
Agricultural Development and Advisory Services, Staplelake Mount, Starcross, Exeter, UK

A. John F. Webster
Department of Animal Husbandry, Bristol University, Langford House, Langford, Bristol, UK

Index